Post Harvest Handling and Storage of Seeds

NIPA® GENX ELECTRONIC RESOURCES & SOLUTIONS P. LTD.
New Delhi-110 034

About the Editors

Dr. Sharmila Dutta Deka has more than 32 years of experience as a Principal Seed Scientist whose skills and expertise are primarily based on improving seed quality characters. Dr. Sharmila has worked on a research project collaborately with International Rice Research Institute (IRRI) in improving the aromatic glutinous joha rice varieties of North eastern regions of India. Dr. Sharmila has skills and experience in conservation of community seed banks and also a member of national monitoring team of AICRP Seeds. Documentation of local germplasm has been done by Dr. Sharmila and farmer's varieties were sent to PVPFR (Plant Variety Protection and Farmer's right) Authority of India for registration and protection of farmer's right.

Dr. Priyanka Sharma, currently working as an Assistant Professor, Assam Agricultural University, Jorhat, India has obtained her Bachelor's degree in Agriculture and Masters in Agriculture with specialization in Seed Science and Technology from Assam Agricultural University, Jorhat. She pursued her Ph. D degree in Seed Science and Technology from Uttar Banga Krishi Vishwavidyalaya, Pundibari Cooch Behar West Bengal India. She qualified ASRB-ICAR-NET and has published several research papers, review articles and popular articles in national and international reputed journals. She is the recipient of Senior Research Fellowship under Department of Biotechnology, New Delhi, Govt. of India. In addition, she has also published several book chapters related to medicinal plants. Besides carrying her research works on quality seed production, she has extended her research interests in the areas of Plant Biotechnology. She was involved in teaching as an Assistant Professor Beehive College of Management and Technology, Dehradun, Uttarakhand. She is actively involved in teaching, research in addition to conducting training programs and workshops related to entrepreneurship development in terms of production of quality seeds and planting materials in medicinal and aromatic plants and other agricultural and horticultural crops.

Post Harvest Handling and Storage of Seeds

Sharmila Dutta Deka
Principal Scientist
AICRP on Seeds, Department of Plant Breeding & Genetics
Assam Agricultural University
Jorhat-785 013, Assam

Priyanka Sharma
Assistant Professor
Seed Science & Technology Programme
Department of Plant Breeding & Genetics
Assam Agricultural University
Jorhat -785 013, Assam

NIPA® GENX ELECTRONIC RESOURCES & SOLUTIONS P. LTD.
New Delhi-110 034

NIPA• GENX ELECTRONIC RESOURCES & SOLUTIONS P. LTD.

101,103, Vikas Surya Plaza, CU Block
L.S.C.Market, Pitam Pura, New Delhi-110 034
Ph : +91 11 4386 0225, 9717133558, 9540816132
E-mail: newindiapublishingagency@gmail.com
Website: www. niparesources.com

Print ISBN: 978-93-58873-47-4

ebook ISBN: 978-93-58874-51-8

Composed and Designed by NIPA®.

Preface

This book entitled "Post Harvest Handling and Storage of Seeds" will cover the topics in detail in terms of seed drying, seed processing, storage, in addition to seed de-linting, seed blending, physical, physiological and chemical changes during seed storage, seed extraction of vegetable crops packaging, seed storage insects and pests and its management. Here are the glimpses of the contents that has been divided into several parts which will be included in this volume of the book.

Seed processing plays a crucial role in modern agriculture by ensuring the availability of high-quality seeds that contribute to increased crop productivity, improved resilience to environmental stress, and enhanced food security. Seed storage structures play a crucial role in preserving the genetic diversity of plant species, ensuring food security, and supporting agricultural sustainability. Seed grading involves sorting seeds based on various characteristics such as size, shape, weight, purity, and germination rate. The grading techniques plays a crucial role in ensuring the quality, uniformity, and viability of seeds, ultimately contributing to successful crop production and agricultural sustainability. The principles of seed storage are crucial for maintaining the viability and quality of seeds over time. It helps in preservation of genetic diversity, maintenance of viability, conservation of rare and endangered species etc. Assessing physical, physiological, and biochemical changes during seed storage is fundamental for maintaining seed viability, ensuring seed quality, predicting longevity, optimizing storage conditions, preserving genetic diversity, ensuring quality control, and driving research and development efforts in seed conservation and management. Seed drying is a crucial process in agriculture and seed preservation, aimed at reducing the moisture content of seeds to a level that minimizes the risk of microbial growth and seed deterioration. Seed de-linting is a crucial step in the seed processing industry, especially for cotton seeds. By de-linting seeds and assessing mechanical damage, seed producers can ensure the delivery of high-quality seeds to farmers, thereby maximizing crop yield and profitability. Packaging plays a crucial role in maintaining seed quality and longevity, as well as facilitating proper handling and storage. Selecting appropriate packaging materials is essential for preserving seed quality, longevity, and handling efficiency. Seed extraction is a crucial step in vegetable crop production, playing a significant role in ensuring the quality, purity, and availability of seeds for future planting. Seed blending play a crucial role in ensuring uniformity, managing risks, optimizing traits, providing flexibility, reducing costs, assuring quality, complying with regulations, and accessing markets. By adhering to these principles, seed producers and farmers can enhance the efficiency, productivity, and sustainability of agricultural systems. Pre-storage seed treatments are essential for seed invigoration, as they play a crucial role in maintaining seed viability and vigour during storage. It also optimizes seed quality, protect against deterioration, and enhance germination and crop performance. Classification of stored pests and insects is crucial for the effective management, preservation, and quality assurance of cereals, pulses, and oilseed crops. Insect and pest infestations pose significant threats to stored commodities, affecting their quality, quantity, and market value. IPM emphasizes preventive measures, monitoring pest populations, and using the least harmful methods to control infestations while minimizing environmental impact.

This book will not only cover topics for undergraduate and post graduate students, based on ICAR syllabus, but it will also be helpful for the stakeholders, policy makers etc. In addition, the readers would also be benefitted by the resources that is available in the book for further getting detailed information with regards to processing, storage, packaging, seed blending, storage insects pests and its management etc.

Authors

Contents

List of Contributors

Dr. Priyanka Sharma, Assistant Professor, Seed Science and Technology Programme Department of Plant Breeding and Genetics, Assam Agricultural University, Jorhat-785013 Assam, India

Dr. Sharmila Dutta Deka, Principal Scientist, AICRP on Seeds, Department of Plant Breeding and Genetics, Assam Agricultural University, Jorhat-785013, Assam, India

Dr. Abhilisa Mudoi, Junior Scientist, Entomology, AICRP (Seeds), Assam Agricultural University, Jorhat-785013 Assam, India

Ms. Meghali Barua, Assistant Professor, Department of Plant Breeding and Genetics, Assam Agrcultural University, Jorhat-785013

Dr. Aradhana Phukan, Assistant Professor, College of Horticulture and Farming System Research, Nalbari Department of Basic Science, Assam Agricultural University, Jorhat-785013

1

Design of Seed Processing Plant-Equipments, Estimation of Processing Efficiency and its Application in Agricultural Crops

Priyanka Sharma and Sharmila Dutta Deka

Concept: Operations conducted before and after harvest are essential components of the system for producing high-quality seeds. If enough care is not taken in the initial stage, there most likely won't be any resources left to get high-quality seed. Like this, carefully produced, high-quality seed may lose a significant percentage of its worth if post-harvest management protocols are not adhered to. Its quality is highly correlated with soundness, fullness, and homogeneity of physical characteristics (size, shape, density, colour, and appendages), in addition to mechanical planting's simplicity and improved crop establishment. Improving the seed that has been gathered and getting it ready for secure storage till needed are accomplished through the effective and efficient process of seed processing. To increase the quality of the seed lot, the goal of seed processing must be to optimize the physical properties of the seed and the constituents of the harvested seed mass. The three main steps in seed processing are conditioning, grading, and applying protectants or enriching with nutrients. These steps are completed in compliance with established protocols and standards (McDonald and Copeland 1997; Agrawal 1996; Jorgensen and Stevens, 2004). The distinct qualities of the crops and the surroundings are taken into consideration while adjusting the machines and operations.

Conditioning: Conditioning allows for the safe handling of raw seed lots during following seed processing activities and maximizes the efficiency of related machinery. To make the lot free-flowing and free from mechanical abuses during later processing steps, it may involve dockage removal, appendage removal, moisture conditioning, or a combination of these.

Moisture contents under 10% are generally regarded as safe for mechanical processing. Excessive or insufficient moisture content can cause mechanical harm, particularly to the heart. To maintain a safe moisture content before undergoing additional processing processes, it is advised that seed batches must be dried or humidified.

Moisture Conditioning: If the moisture level of the raw seed lots is too high, it may be required to dry them, and if it is too low, to humidify them before preparing them for safe mechanical handling.

Tempering: In addition to drying, many crop seeds (such those of soybeans and peas) require tempering in extremely dry conditions to get the moisture content up to roughly 10%. Tempering is necessary for crops that are vulnerable to impact damage at lower moisture content levels. Raising the moisture content to a safe range of roughly 10–12% can be accomplished through tempering.

Removal of Appendages: Seeds with awns, beards, hairs, glumes, and other appendages tend to interlock and cluster together. To enhance the flow qualities, cleaning capabilities, and seed quality, these must be eliminated. The pebble mill, debearder, and hammer mill are the common equipment utilized in this process.

Hammer Mill: It is a thresher with concave, oscillating sieves, an aspirating blower, and hammer-style beaters in a closed cylinder casing. A hammer mill can grind grasses like blue bunch wheatgrass and blue wildrye, and species with similar seeds like tall oat grass, bulbous barley, and squirrel tail, satisfactorily (Sinha and Srivastava, 2003). Pusa Basmati 1, a well-known long-awned paddy variety, has a problem with mechanical processing because its awns get tangled. To prepare the seed mass for mechanical processing, de-awing is accomplished with the hammer mill working at ideal settings by adjusting the feed rate, screen opening, and cylinder speed (Sinha *et al.* 2010).

Debearder: Inside a steel drum, there is a horizontal beater with revolving arms. To transfer the seeds, the arms pitch through the drum. Inside the drum are stationary pillars that can be adjusted to provide for clearance with the arms. Both the arms and one another are massaged with the seeds. The processing time can be adjusted by adjusting the discharge gate. The beater speed, beater clearance, and processing time all affect the degree of action. It has more capacity than a hammer mill, is easier to use, and causes less harm to the seeds. Clusters of grass seeds, awns, beards, and hulls from some grass seeds, clip seed from oat seed, debeard barley seed, thresh white caps in wheat, and cotton webbings from Merion bluegrass may all be done with this equipment. It can also be used to polish the grass seeds to improve their quality.

Pebble Mill: Its drum revolves around a shaft that is installed at each end, off centre. The mill is loaded with smooth, half-inch pebbles and is turned slowly until the rubbing action of the pebbles causes the fuzz from the seeds to roll into little, round balls. The mixture of pebbles, seeds, and matted fuzz is then run over a scalpel to remove the stones. When it comes to eliminating fuzz and seed hairs, such as the cobwebby hairs from seeds like bluegrass, these kinds of devices are quite efficient.

Scalper: From a thresher or combine, seeds that have been subjected to more debris, leaves, weed seeds, chopped grains, and infected seeds may be transferred to a cleaning plant. These compounds make it impossible for cleaning or separation devices to handle seed mass effectively. Scalpers are typically used to remove it. In general, there are two types of scalpers that are used: (a) a metal screen reel with holes that is rotated around a central shaft and is slightly tilted; (b) seeds that are fed into the higher end of the reel tumble inside it until they fall through the holes. Longer trash, on the other hand, is released separately and stays in the reel. The scalper and pre-cleaner air screen equipment are integrated in small capacity facilities.

Huller and Scarifier: From a thresher or combine, seeds may be transferred to a cleaning facility that have been subjected to more dirt, leaves, weed seeds, chopped grains, and contaminated seeds. These compounds make it impossible for cleaning or separation devices to handle seed mass effectively. Scalpers are typically used to remove it. Two kinds of scalpers are used: (a) a slightly slanted reel of metal screen with perforations that revolves around a central shaft, and (b) seeds fed into the higher end of the reel tumble inside it until they fall through the perforations. Longer trash, on the other hand, is released separately and stays in the reel. The scalper and pre-cleaner air screen equipment are integrated in small capacity facilities. Commercial scarifiers work by pressing the seeds up against an abrasive substance, like carborundum, to scarify them. The seeds can be brought into contact with the abrasive surface using either a centrifugal force or an air stream. Certain scarifiers, particularly those for delicate seeds, employ a rubber abrading surface in place of carborundum to avoid harm. High moisture content seeds are challenging to hull or scarify since the huller or scarifier will deteriorate the seeds. After being hulled and scarified, seeds that remain viable for a long time can be processed right away after harvest and kept for the next growing season. Some that quickly lose their viability can be preserved, hulled, and scarified only prior to planting season. Depending on if hard seed, unhulled seed, or both are present, hulling and scarification can be done individually or together. Some seeds, like those of buffalo grass, Bahia grass, and Bermuda grass, just need to be hulled; however, some seeds, such as those of suckling clover, alfalfa, crotalaria, hairy indigo, subclover, and wild winter peas, might also require effort. But some seeds, notably sour clover, black medic, crown vetch, sweet clover, and *Sericea lespedeza*, may require scarification in addition to hulling (Sinha and Srivastava, 2003).

Removal of Dockage: Every now and then, more than 20% of the harvested seed mass is dockage by volume. Mechanical processing of such lots is not recommended since it will increase seed loss through the reject port and decrease processing equipment efficiency. Under these conditions, the initial step for such lots is winnowing, which separates them from the seed lot by taking use of the dockages' varying aerodynamic behaviours. For this purpose, it is recommended to employ an adjustable air flow in conjunction with a human safety winnower. There are numerous commercially available varieties of winnowers.

Cleaning and Grading: It is necessary to eliminate any significant amounts of moderate contaminants, such as plant fragments, broken kernels, dirt clods, dust, stone, etc., from the harvested seed mass. Apart from these impurities, the seed mass comprises seeds with varying attributes such as size, shape, density, colour, texture, and others, all of which diminish the quality of the seed in terms of its biological and commercial value. Cleaning and grading are required to enhance the biological and physical quality of seeds using the following units.

Air Screen Cleaner Cum Grader: By removing tiny contaminants and undersized seeds from the seed lot, it enhances both the physical purity of the seed lot and the material flow in later processes using screens and air streams (aspiration) (sieves). Essentially, the process of cleaning involves utilizing the distinction in aerodynamic properties between the seed and the impurities present in the seed lot, with the thickness of the seed serving as a grade element. It's crucial to remember that the thickness alone determines the grade, not the length of the seed. A system of aspiration is also incorporated into the machine, providing air streams at two different points: (a) prior to the suction channel, which removes lighter impurities prior to the seed mass reaching the screens; and (b) following the suction channel, which removes impurities and lighter seed components that remain after the seed mass has been cleared. The Air Screen Cleaner cum Grader uses two sets of screens for grading and scalping. Although the cleaning efficiency rises with more screen sets, the throughput capacity falls. The machine's performance is determined by its operational characteristics, which include feed rate, oscillation speed, screen pitch, air blast volume, and selection screen size and type. To pass materials other than excellent seeds and ride over all good seeds, the grading (bottom) screen should be narrower than the optimal thickness/diameter of good seeds. To pass all of the seed components and scalp off the coarse contaminants, the scalping (top) screen needs to be big enough. As shown in Table 1, consideration should be given to the size and form of the seed to be processed when selecting the type of screen.

Nonetheless, the basis for selecting the screens' opening size will be the sieve analysis for that specific seed lot as well as the crop seed's appropriate requirements for grading and scalping screens. According to Trivedi and Gunasekaran (2013), Table 2 shows the screen requirements for processing key significant crop species in India. The screen cradle's oscillation speed, pitch, and slope should be adjusted to allow the seed to tumble and spin on its own axis vertically while being graded by a screen opening that has an efficient cleaning and grading system and a high processing throughput. The scalping screen is set to a high slope for rapid removal of debris and weed seeds, while the grading screen is set to a flat slope to hold the seed longer for close separation. For spherical seeds, the pitch should be flattened; otherwise, it will bounce over the screen and reduce grading efficiency. For this reason, chaffy seeds should be steeped in the pitch. The flow velocity of the seeds over the screens must be properly controlled to provide improved cleaning and grading. It needs to be modified based on the seed's form and mass dockage level. Higher dockage seed mass should be maintained with a lower feed rate. Feed rate needs to be adjusted to maximize the amount of time the three-layer seed stays on the screen. To maximize efficiency, the machine's screen cleaning mechanism is also crucial. Air screen machines use three different methods of screen cleaning: (a) brushes, (b) rubber balls, and (c) knockers or beaters. The rubber ball method is the most effective of all since it reduces the likelihood of mechanically mixing seed and causes the least amount of wear and tear on the screen. While the structural strength of the screen decreases with increasing screen length, screening efficiency rises. The flow velocity of the seeds over the screens must be properly controlled to provide improved cleaning and grading. It needs to be modified based on the seed's form and mass dockage level. Thus, two Air Screen Cleaner cum Graders are utilized in seed processing to increase the machine's service life as well as the cleaning and grading efficiency. The first machine's main function is cleaning, whereas the second one's main function is grading. Remarkably, the air screen machines are also capable of efficiently eliminating dockages and dead seeds (Sinha *et al.* 2001; Sinha *et al.* 2002).

Table 1: Selection of the right kind of screen based on the size and form of the seed to be processed.

Shape of seed	**Screen**	**Opening of screen**
Round	Scalping	Round hole
	Grading	Slotted hole
Oblong	Scalping	Oblong hole
	Grading	Oblong hole
Lens shaped seeds	Scalping	Oblong hole
	Grading	Round hole

Table 2. Size of screen apertures for processing of major field crops

Screen aperture sizes in millimetres		
Crop	**Top screen**	**Bottom screen**
Paddy		
Coarse grain/bold type	2.8s, 9.0r (3.2s)*	1.85s (2.1s)*
Medium slender	2.8s, 9.0r	1.80s
Fine/Superfine	2.8s, 9.0r (3.2s)*	1.70s (1.8 or 2.1s)*
Wheat		
Triticum aestivum	6.00r (5.5r)*	1.80s, 2.10s, 2.30s (2.3 or 2.4s)*
Rapeseed and Mustard		
Mustard	2.75r, 3.00r, 3.25r, (3.25r)*	0.90s, 1.00s, 1.10s, 1.40r (1.6 or 1.8r)*
Maize		
Maize except popcorn	10.50r, 11.00r (10.50r)*	6.40r, 7.00r (6.0s or 6.40s+8.00r)*

r- round, S-slotted apertures

(Source: Trivedi & Gunasekaran, 2013)

Indented Cylinder Separator: It is also called a length separator because it uses the length difference between the seed and contaminants of the seed mass that are either longer or shorter than the crop seed, especially lengthwise broken seeds. The movable horizontal separating trough positioned inside a revolving, almost horizontal cylinder makes up the indented cylinder separator. The word refers to the small, closely spaced semi-spherical indentations on the cylinder's inside surface that resemble cells or pockets.

The fed seed mass passes through the bottom of the cylinder from the feed end to the discharge end. The cylinder rotates, rotating the seed mass so that it may be measured by the cylinder's recessed indents. The lifting trough is used to collect short seeds and/or impurities that have been removed from the seed mass. Long seeds stay inside the cylinder and come out through a different spout at the end. The shorter impurities, which have been raised to a specific height by the cylinder's rotation motion and recessed indents, fall into the separating trough when gravity pulls stronger than the centrifugal force produced by rotation. The seed's ability to fit into the indent and be dragged out of the seed mass is influenced by its length, centre of gravity, surface texture, and size. The efficiency of the separation is influenced by cylinder speed, retarder position, feed rate, cylinder pitch, and indent size, among other operational aspects. With cylinder speed, the centrifugal force holding the seed in the indents increases, lifting the seed longer or higher and facilitating its separation. The seed will, however, fall out of the low-level indents if the cylinder speed is decreased because doing so will also result in a decrease in centrifugal force. The form and size of the indent must be selected to best suit the requirements of that specific seed batch. To ensure that the material caught in the indents falls into the trough, the trough's position must also be adjusted. For fine separation, the retarder setting, and the cylinder's longitudinal tilt or pitch should be adjusted.

Specific Gravity Separator: A gravity separator removes undesirable seed components or contaminants from the seed mass that is processed by taking advantage of the differences in density between high-quality and low-quality seeds and trash. If the cleaned and graded seed mass (using an air screen and an indented cylinder) still contains high-density material such as clods, stones, or lower density grain components (immature, insect-damaged, or diseased seed) that were not separated by previous operations, the seed mass needs to be processed for density grading. Using the floating principle, the gadget arranges seeds vertically in layers on the deck according to their density. Distinct density grain components or lot impurities follow distinct routes due to deck vibration and tilt. The specific gravity separator is fed with layers of seeds that range in thickness from three to five. The heavier seeds are propelled upwards by the deck's trembling and the deck's differential air stream, while the lighter seeds descend. Several movable grates that are used to sort materials according to their densities comprise the machine. Clumps or stones are forced to the transverse edge, where materials are sent to various outputs via movable grates. The air volume, feed rate, oscillation speed, side slope, and end slope are the operational factors that have a direct impact on the machine's performance. The seed mass on the deck is stratified using a differential air volume adjustment from the feed to the discharge end. The heavier seeds should be placed on the deck and the lighter ones should be raised into the top layer of the seed mass by adjusting it accordingly. The air volume on the feed side should be greater than the air volume on the outlet side.

The rate at which the seed mass travels over the deck is determined by the slope of the deck from the feed hopper to the discharge end. The flat end slope is required to maintain the seed mass on the check for an extended amount of time when there is minimal difference between the seed and impurities or the seed that needs to be separated. The ultimate slope may, however, rise if there are notable differences in the specific gravities of the contaminants and seed. Light seed layers glide downhill to the lower side of the cheek on an air cushion, while heavy seed is pushed uphill to the other side of the deck oscillation. While processing will inevitably result in some seed loss in the reject port, this loss can be greatly minimized by operating the machine at the best possible settings for the type and quantity of seed being processed. Specific gravity separation has been shown to result in seed loss of between 1 and 3%, with lower seed lots experiencing greater seed loss (Sinha *et al.* 2002).

The process of separation yields a range of seeds, from light to heavy. To obtain high-quality separation, operators must have experience and practice making the following adjustments

a) Feed rate: The volume of seed that is placed on the gravity table. The processing line's capacity determines the rate. To guarantee consistency and continuity, place a slide beneath the above feed hopper.

b) Side tilt: The variation in elevation between the deck's high and low points.

c) End raise: The gravity table's gradient from the feed to the discharge terminates. The slope affects how long the seed remains on the surface for cleaning.

d) Ecentric speed: The oscillation speed of the gravity table.

Pneumatic Separator: It takes advantage of the variations in the terminal velocities of the impurity and seed air streams. Physical characteristics that impact a particle's resistance to air flow and variation in terminal velocity include density, shape, and surface roughness. The limited rising air stream receives the seed mass and raises all the particles to varying levels based on their terminal velocities. Different fractions contain particles with varying terminal velocities. The pneumatic separator's air system forces air through the apparatus by exerting pressure that is greater than atmospheric pressure. As opposed to this, the aspirator's fan, which is located at the discharge end, creates a vacuum that enables air pressure to push through the separator. Its components are as follows: a control unit that modifies the pneumatic pressure based on the amount of seed mass that requires processing.

Electromagnetic Separator: It divides the components of the seed mass according to variations in their electrical characteristics. The degree of separation is determined by the relative ability of the seeds in the mixture to conduct electricity or hold a surface charge. The belt receives a surface charge from the application of a thin layer of seed onto it *via* a high-voltage electrical field. This charge is comparable to that which a comb runs through hair to pick up. When a belt circles a pulley, seeds that have lost their charge fall off the belt generally right away. Poor conductor seeds, which lose their charge slowly, will stick to the belt and eventually fall off. After then, the dividers in the drop path can be positioned to gather any desired portion of the distribution. Feed rate, belt speed, electrode location, voltage, and divider position are the machine's operating parameters. All modifications are related to the seed's moisture content since it influences the electrical conductivity. Watercress seeds from rice, ergot from bent grass, and Johnson grass from sesame can all be successfully separated with electronic separators.

Spiral Separator: Seed mass is categorized based on the morphology of the particles, which is connected to rolling resistance. It is made up of spiral-shaped sheet metal strips that are arranged around a central axis. The apparatus has a vertical, open screw conveyor-like appearance. At the apex of the inner spiral, the seed is fed. Because they have less rolling resistance than flat or irregularly shaped seeds, round seeds roll more quickly. The round seed's orbit speeds up as it revolves around the axis and rolls over the edge of the inner flight into the outer flight, where it is gathered separately. Conversely, the unevenly shaped or flattened seed slides or tumbles, moving slowly and failing to gather enough momentum to break free from the inner flight. As a result, each is gathered independently from the inner flight outlet at the bottom. When it comes to separating damaged seed from peas, lentils, brassicas, vetch, and soybeans, the spiral separator is incredibly helpful. Spiral separators are often helpful in eliminating fragmented grains, seeds with varying textures, and contaminants dispersed within the seed mass. Sinha *et al.* (2008) state that it is also appropriate to separate seeds that are damaged or round (in terms of shape) from a mixture of damaged seeds that are round or flat.

Inclined Draper: It divides the materials according to variations in surface roughness, specific gravity, and rolling or sliding characteristics. The mixture is uniformly distributed throughout the middle area of a flat surface coated in plastic, velvet, or slanted velvet that is adjustable, help of a hopper that feeds the seed mass. Smooth or spherical seeds travel to the lower end of the belt as it advances upwards, whereas materials with rough surfaces, oblong shapes, or flat surfaces stay on the belt. To maximize the effectiveness of separation, the belt's angle, speed, and seed flow rate can all be modified based on the characteristics of the seed surface. Higher capacity processing can be achieved by using multiple inclined drapers. Cleaning flower seeds, separating fruits, seed clusters, and plant debris, and extracting buckhorn plantain (*Plantago lanceolata*) from red clover (*Trifolium pratense*) are all made easier with this handy tool.

Spiral Separator: Seed mass is categorized based on the morphology of the particles, which is connected to rolling resistance. It is made up of spiral-shaped sheet metal strips that are arranged around a central axis. The apparatus has a vertical, open screw conveyor-like appearance. At the apex of the inner spiral, the seed is fed. Because they have less rolling resistance than flat or irregularly shaped seeds, round seeds roll more quickly. The round seed's orbit speeds up as it revolves around the axis and rolls over the edge of the inner flight into the outer flight, where it is gathered separately. Conversely, the unevenly shaped or flattened seed slides or tumbles, moving slowly and failing to gather enough momentum to break free from the inner flight. As a result, each is gathered independently from the inner flight outlet at the bottom. When it comes to separating damaged seed from peas, lentils, brassicas, vetch, and soybeans, the spiral separator is incredibly helpful. Spiral separators are often helpful in eliminating fragmented grains, seeds with varying textures, and contaminants dispersed within the seed mass. Sinha *et al.* (2008) stated that it is also appropriate to separate seeds that are damaged or round (in terms of shape) from a mixture of damaged seeds that are round or flat.

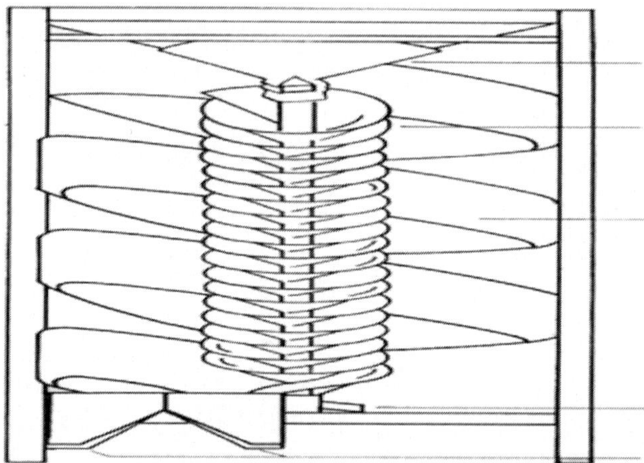

Magnetic Separator: It also takes use of the variations in the seed's surface texture. Iron fillings are applied to the seed lot; smooth seeds do not pick up the fillings, whereas rough seeds do. A rotating magnetic drum is passed over the treated seed. The magnetic material-containing seeds are drawn to the drum, fall close to it, or stay stuck to it. The elements that cling are removed at the conclusion of a batch. The other elements separate as they descend closer to the drum. According to Sinha and Srivastava (2003), Dodder (*Cuscuta pentagona*) from clover and alfalfa, red clover, and *Sinapis arvensis* (wild mustard) from brassicas are all extracted using a magnetic separator. *Stellaria media* (chickweed) is also extracted from clover and alfalfa.

Horizontal Disk Separator: To separate distinct fractions, it makes use of centrifugal force, rollability features, and differences in form and surface roughness. Seeds are confined to the centre of a flat rotating disk by a stationary circular plastic fence, with seeds being metered to the outer portion of the disk through adjustable outlets. Centrifugal force causes round or smooth seeds to roll or glide off the disk. Rough or irregular seeds are transferred into a separate hopper and stay on the disk. To ensure that every seed travels separately, the feed rate through the outlets must be changed. Like the spiral separator, the horizontal disk features a disk speed control that allows it to alter the percentage of seed that is thrown off or retained, making it more selective. To improve capacity, many disks can be mounted on a single vertical shaft. When one seed has a larger inclination to roll or slide than another, the horizontal disk separator can be used to separate curly dock from red clover, dodder seeds from lucerne (alfalfa), and other mixes.

Velvet roller: Seeds with similar sizes but different levels of roughness are separated by the velvet roller. It is helpful for fodder species, especially clover, whose blooming pattern variation causes uneven seed development at harvest and wrinkles a lot of the seeds when they dry. The machine is composed of two parallel cylinders that are joined together and covered with a unique natural fibre material. At a speed of 0.5 to 1 revolution per second, the cylinders spin in opposing directions with a minor inclination. The seed descends the gradient by passing between the two revolving cylinders. Sharp-edged or rough-surfaced seeds are lifted by the rotational movement and fall into a collecting bin. Sections of the collecting system are used to separate seeds with varying levels of surface roughness.

Colour Sorting machine: When other physical characteristics like thickness, length, density, and surface roughness are unable to distinguish between discoloured and good seed lots, the colour sorter is mostly utilized to do so. Seed discoloration can be caused by mechanical damage to the seed coat, pest infection, or unfavourable weather. The device compares the backdrop filter's colour with the seed colour using photoelectric cells or sensors. The backdrop filter is chosen to reflect light with a wavelength that is comparable to that of high-quality seeds. A vibratory feeder measures the amounts of seed to be colour-sorted, feeding one seed file at a time into the sorting chamber. When photoelectric sensors detect a difference in colour in a seed, they measure the reflected wavelength of light from the discoloured seed. This information is then used to create an electric impulse and launch an air jet to expel the discoloured seed from its regular course. The colour sorter is generally used to separate discoloured and good seed lots when other physical attributes such as thickness, length, density, and surface roughness are not able to do so. Seed discoloration can be caused by mechanical damage to the seed coat, pest infection, or unfavourable weather. The device compares the backdrop filter's colour with the seed colour using photoelectric cells or sensors. The backdrop filter is chosen to reflect light with a wavelength that is comparable to that of high-quality seeds. A vibratory feeder measures the amounts of seed to be colour-sorted, feeding one seed file at a time into the sorting chamber. The good seed continues the typical course, and it is differentiated from the discoloured seed because it has veered from its usual feeding pattern. For a seed lot, the colour sorter should only be utilized after the fundamental grading and cleaning. The machine's operating characteristics, such as the vibration of the feeder, the speed of the feeding belt, the feed rate, the location of the discharge point, the background colour filter, the colour range, the ejector timings, and the ejector lag time, all have an impact on the machine's efficiency. Significant improvements are made to the lot's purity, germination percentage, Vigor index, and real density by the electronic colour sorting. But only when seed batches have a colour purity level of at least 70% is the colour separation technique effective (Sinha and Modi 2000; Sinha and Vishawkarma, 2001).

X-Ray Sorter: A combination of radiography and image processing is used to process seed lots based on their physical attributes, including size, shape, density, colour, texture, and individual seed anatomy. Using non-invasive X-ray-based imaging, the inner architecture of treated, coated, or pelleted seed can be inspected. After being grouped together in a specific holder and subjected to long-wave X-rays, individual marked seeds are digitally radiographed. The radiograph undergoes additional conversion to a 3D image, after which it is processed digitally. Digital analysis and comparison are performed on the processed image data using the quality seed database's digital data as a reference. It is crucial to remember that internal morphology, not biological value, is the basis for X-ray imaging and sorting.

Other Basic Seed Equipment in the Processing Plant

The components of the seed cleaning plant include the holding bins, dust collection systems, bucket elevator, intake/reception hopper, and seed treater. They are linked together directly or through ducting. The national/city power grid or a diesel generator provide the electricity needed to run the complete system. Any machine's operator or technician must understand how each component works, be familiar with all of its elements, and be able to make the necessary adjustments to ensure optimal performance.

1) **Feed or intake hoppers:** The container into which the uncooked seed is emptied is called the intake hopper, often known as the "reception pit" when it is buried in the ground. It has an overlaying steel grill and is constructed of concrete or steel. For the seed to flow by gravity into the conveyor from the hopper, the sides of the hopper must be at an angle of at least 35°. The flow is managed using the capacity regulation slide. Sections of ducting link the bucket elevator's intake hopper to the input hopper. Seeds can pass through the opening in the steel grill, but big items like straw, stones, and other foreign objects stay inside. Additionally, the steel grill makes sure that no empty sacks get stuck in the equipment (Kugbei *et al.*, 2017).

2) **Bucket elevator:** Seed is vertically lifted into other processing line machinery by the bucket elevator. It is divided into three parts:

   a) Boot - lower portion: a steel shell, tension device, inflow hopper, capacity adjustment slide, and pulley with shaft housed in bearings. Bolts are used to fasten the boot to the ground. Typically, a sliding screen at the bottom can be taken out for cleaning to eliminate any debris.

   a) Legs - middle portion: typically self-supporting, positioned on the boot, and has a rectangular cross section. It is possible to install inspection windows and connect the dust collection system to certain parts of the legs.

   b) Head - upper section: positioned above the legs, it consists of a steel casing, a drive pulley with a shaft positioned in bearings, a bracket for the power and motor transfer setup, a throat plate, and a discharge chute. A rubberized belt travels continually between the legs and over the two pulleys within the bucket elevator container. Buckets or cups composed of pressed, welded, or moulded plastic are fastened to the belt using special nuts and bolts.

   c) The bucket elevator can sustain itself. Nevertheless, to guarantee that the elevator remains vertical and sturdy while in use, brackets are frequently fastened to the floors, walls, and ceiling.

3) **Holding bins** – Buffer bins, surge bins, and bagging bins regulate the seed flow of the conveying and handling system. In front of every bin is the proper piece of equipment (e.g., bagging scales, in-line process

scales, cleaners, gravity separators, and seed treaters). Hopper bins usually have a rectangular cross-section and a self-emptying hopper at the bottom that is angled less than 45 degrees. Sight glasses, slide gates, level sensors, and flanged air vent connections to cloth filters or central dust collection systems are features that certain bins include (Kugbei *et al.*, 2017).

4) **Ducting:** Typically, it takes the shape of elbows, junction boxes, diverters, bends, and straight pipe segments. For seed to flow by gravity, ducting should be sloped between 35 and 45 degrees. Bolts, clamps, or clips are used to assemble the ducting components. Certain ducts include windows, sliding lids, or detachable lids installed to allow for processing-related inspection. The processing line's numerous equipment components are connected by pipework made up of ducts, chutes, and spouts (Kugbei *et al.*, 2017). Two fundamental layouts exist:

    a) A machine stacked on top of another, with seed moving through small connections due to gravity.

    b) Equipment on the same floor, with ducting connecting the machines and bucket elevators transporting seed from one to the next.

Processing Equipments used for Improving the Quality of the Seed

Depending on their specificity and specialization, the seeds must go through a variety of seed processing equipment stages from harvest to the last stage of seed storage. However, some tools, such as graders and seed cleaners, are universal for all kinds of seed. The following list includes the processing tools and equipment used in handling seeds.

S.No.	Processing apparatus	Application in relation to particular seed management
A.	**Using extraction machinery for threshing**	
1.	Thresher	To extract the seeds, particularly from cereals, from the inflorescence
2.	Ginning machine	To extract cotton's seed and lint from the kapas
3.	Maize sheller	Removing the seed shell from the cobs
4.	Pulse thresher	To remove the seeds out of the pods
5.	Tomato seed extractor	To remove the tomato seed from the fruit without throwing away its pulp
6.	Chilli seed extractor	For simple seed extraction from chili fruits
7.	Groundnut decorticator	The kernel's shell (seed from the pods)
8.	Sunflower thresher	For removal of the seeds out of the head
9.	Debearder	To extract the awns from the seed of barley
10.	Mechanical scarifier	Mechanically scarifying hard seeds in order to increase seed germination

S.No.	Processing apparatus	Application in relation to particular seed management
11.	Pebble mill	To remove grasses of their hairy webs
12.	Timothy bumper mill	To extract the weed seed from the Timothy seed
13.	Hammer mill	Removing the seeds' hook or other appendages (e.g., Stylosanthus)
B.	**Driers**	To get the moisture content down to the minimum required for safe handling during the final stages of processing and storage
C.	**Grading equipments**	
14.	Cleaner cum grader	This process, called fundamental grading in seeds, homogenizes the previously cleaned seed according to size. The required sizes of the sieves vary by crop.
15.	Precleaner and aspirator	By doing this, the inert material and dust particles are removed from the seed, increasing the grading efficiency.
D.	**Upgrading machines**	
16.	Specific gravity separator	By utilizing its weight or specific gravity, graded seed can be further improved in quality. Larger seeds indicate maximum field establishment and are good keepers.
17.	Indent cylinder	It keeps longer seeds the same size (both in length and width). Good seeds are chosen after the broken and damaged ones are removed.
18.	Disc separator	It is done to get rid of weed seeds and make the seed look better overall.
19.	Roll mill	To distinguish between smooth and rough seeds according to their surface texture, particularly weed seeds
20.	Magnetic separator	Elimination of weed seeds from vetch, trefoils, alfalfa, and clovers
21.	Inclined draper	Separating round or smooth seeds from flat, rough, or elongated seeds
22.	Electronic colour sorter	Sorting out discoloured seeds
23.	Electrostatic Separator	Johnson grass is extracted from sesamum seeds based on their electrical characteristics (specific utility)
24.	Spiral separator	Sorting seeds according to their shapes, etc. separation of soybean seed, vetch, and rape from rye, wheat, or oat grass
25.	Polishers	To make the seed more lustrous
26.	Picker belts	To remove unwanted pods or ears from shelled seeds, for example corn, and groundnuts
27.	Vibratory separator	Elimination of weed seeds
28.	Seed treater	Applying fungicide and insecticide to the seed
29.	Seed packing machine	To make the task easier and prevent mixing errors by humans
30.	Conveyers, Elevators (Belts, Buckets)	Makes seed transfer between machines easier and prevents seed contamination at different levels.

(*Source*: www.eagri.org)

References

Agrawal RL (1996) Seed technology. Oxford and IBH Publishing Co., New Delhi, India, p 829.

Jorgensen KR and Stevens R (2004) Seed collection, cleaning, and storage, Chapter. 24. In S. B. Monsen, R. Stevens, and N. Shaw, eds. Restoring western ranges and wildlands. Ft. Collins, CO: USDA Forest Service Gen. Tech. Rep. RMRS-GTR-136

Kugbei S, Ayungana, M., Hugo W. (2017). Training Toolkit Module 2: Seed Processing equipment. The Food and Agriculture Organization of the United Nations and Africa Seeds Rome.

McDonald MB, Copeland LO (1997) Seed production: principles and practices. Chapman and Hall, New York

Sinha JP, Kumar A, Weissmann E (2022). Seed Science and Technology (Biology, Production, Quality). Springer, ISBN 978-981-19-5887-8

Sinha JP, Modi BS (2000) Colour sorting and seed quality in mungbean. J Inst Eng 81:1–5

Sinha JP, Modi BS, Nagar RP, Sinha SN, Vishwakarma MK (2001) Wheat seed processing and quality improvement. Seed Res (29):17–178.

Sinha JP, Sinha SN, Atwal SS, Seth R (2008) Success story (1997–2008): seed Processing & Storage. Published by IARI Regional Station, Karnal, pp 1–62

Sinha JP, Srivastava AP (2003) Seed Processing. In: Singhal NC (ed) Hybrid seed production in field crops. Kalyani Publishers, India, pp 269–296

Sinha JP, Vishawkarma MK (2001) Effect of electro-mechanical colour sorting on seed quality in pigeonpea (Cajanus cajan) cultivar 'Pusa 33′. Indian J Agril Sci 71(4):287–289

Sinha JP, Vishawkarma MK, Atwal SS, Sinha SN (2002) Quality improvement in rice (Oryza sativa) seed through mechanical seed processing. Indian J Agril Sci 72(11):643–647

Trivedi RK, Gunasekaran M (2013) Indian minimum seed certification standards, The Central Seed Certification Board Department of Agriculture & Co-operation. Ministry of Agriculture, Government of India, New Delhi

www. eagri.org

2

Seed Storage Structures and Prediction of Viability During Storage-Viability Nomograph

Priyanka Sharma and Sharmila Dutta Deka

Concept: The preservation of viable seeds from the moment of collecting until they are needed for sowing is known as storage. Every year, a significant quantity of seeds loses their usefulness due to incorrect storage. Seeds need to be preserved and stored dry. As was previously mentioned, seed moisture content and storage temperature have the biggest effects on how long seeds last. The three main environmental elements that impact seed deterioration and loss of viability are moisture, temperature, and oxygen percentage. For over 90% of species, reducing seed moisture content (MC) to certain thresholds consistently lengthens the seed's lifespan (Roberts 1973). These species are categorized as "orthodox" in terms of the conditions in which they should store seeds; under suitable dry, cool conditions, they typically maintain viability and germinability even after extended periods of storage. Ellis and Roberts' improved viability equations have addressed the quantitative impacts of both drying and cooling for orthodox seeds. The life of the seed is often doubled for every 1% drop in seed moisture (when seed MC fluctuates between 5 and 14%) and every 5 °C drop in storage temperature (between 0 °C and 50 °C) (Harrington 1972). Low seed MC and low temperature are hence fundamentals of traditional seed storage. Careful regulation of these two factors is necessary to achieve optimal storage life. But most seed produced annually just has to be kept from harvest until the following planting season. Depending on the type of seed, it may or may not need to be stored at temperatures and relative humidity above room temperature. Regardless of any modifications made for unique uses, the local climate storage facility needs to have a few essential components (Barre, 1954; Wheeler and Hill, 1957).

Purpose of Seed Storage

Naturally, seeds need to be preserved because there is typically a gap in time between harvest and planting. The seed needs to be stored somewhere throughout this time. Although the primary justification for saving seed is the amount of time between harvest and planting, there are additional factors to consider, particularly when storing seed for a lengthy period. From the time of harvest till planting, the goal of seed storage is to keep the seeds in good physical and physiological condition. Acquiring sufficient plant stands is just as crucial as having robust, healthy plants. Not every seed supplier can sell all the seed they produce for the upcoming planting season. Unsold seeds are frequently "carried over" in storage in preparation for marketing in the next planting season. Some types, variations, and large quantities of seed do not transfer particularly well, which causes issues with seed storage. To avoid having to generate the seed every season, seeds are also purposefully kept in storage for long periods of time. This has been found by Foundation Seed Units and others to be a cost-effective and efficient process for seeds of kinds that have low demand. Certain types of seeds are kept in storage for long stretches of time to increase the percentage and speed of germination by giving the seeds adequate time to emerge from dormancy "naturally." Maintaining a sufficient capacity for germination and emergence is the goal of seed storage, regardless of the motivations behind it. As such, the facilities and practices employed in storage must be focused on achieving this goal.

Essential Elements of Storage Buildings/structures

1. **Protection From Water:** Seeds should never come into touch with rain, snow, ground moisture, or any other source of water as this will increase the moisture content of the seeds. Elevated moisture content in seeds leads to increased respiration, heating, Mold growth, and occasionally even sprouting during storage, all of which diminish the quality of the seed. Seed storage facilities must not have any holes or gaps that allow rain or snow to enter through the roof or sides. Wooden walls and roofing should have their cracks and knot holes repaired. Rubber washers should be installed on all nuts and screws in metal buildings, particularly those located on the roof. Given that seeds are easily absorbed by soil moisture upon contact, a waterproof floor is a must for all seed storage facilities. A moisture barrier should be placed beneath a concrete floor, and wood floors should be raised. Although a metal floor is water-tight by nature, constant moisture will cause it to rust out quickly.

2. **Protection From Contamination:** Each seed lot needs to be maintained free of seeds from other lots since individual seeds of different cultivars of different crops are indistinguishable from one another and because the types of seeds are frequently difficult to separate. Buildings used for storage should be designed to offer the highest level of protection against accidental contamination. There needs to be a different container for every cultivar in bulk storage. Each cultivar's seeds need to be stacked separately for bag storage. Tote boxes can also be used to store seeds. Forklifts can easily move tote boxes and palletized bag stacks. Every bag, tote box, and bin needs to have a clear and legible label.

3. **Protection From Rodents:** Precautions need to be taken to keep mice out of storage buildings since they can ruin vast amounts of seeds if they aren't kept safe. Buildings made of concrete and metal typically offer good rodent protection. Wooden buildings can also be built carefully to keep them out. Rats are kept out of metal bins with tight lids, while rats can be kept out of cloth bags by treating them.

4. **Protection From Insects:** While certain types of seeds do not pose a significant threat from insects, a well-designed storage facility should allow for the complete or partial fumigation of the area at any point to effectively manage insects. Every time a bin is emptied, make sure it's completely cleaned and fumigated to minimize insect infestations. Trash and loose seeds should never be present in areas used to store bags or tote boxes. Maintaining cleanliness helps with bug control by removing areas where insects can breed.

5. **Protection From Fungi:** Certain types of storage fungus have the potential to seriously harm seeds that have been stored. Storage structures should provide cool, dry temperatures, as fungus prefer warm, humid environments to develop. By thoroughly drying seeds to a safe moisture content—typically less than 12 percent—and storing them in dry circumstances, damage from storage fungus can be minimized. Ventilation is sometimes required to keep translocated moisture from building up in storage structures, particularly steel buildings. Water vapor may migrate from warmer to cooler sections of a bin due to temperature differences; this movement usually occurs to the upper surface. If sufficient measures are not taken to avoid its accumulation, such moisture flow may create an environment that is conducive to the growth of fungi. To manage fungus, seeds can be treated with a variety of substances. Many different types of seeds often receive fungicide applications as part of their regular processing. Fungicides used to manage soil fungus might not, however, be able to manage storage fungi.

6. **Protection From Fire:** Wooden buildings are most at risk from fire loss; however, wood can be chemically treated to slow down burning. Maintaining cleanliness inside and outside the building can help lower the risk of a fire in a wood-framed structure. Special dust proof and spark proof electrical outlets and switches should be installed in all seed storage facilities, particularly in processing sections, as they significantly lower the risk of an electrical fire. Every wire needs to be resistant to rodents. Even though concrete and metal structures are fireproof, electric sparks can produce dust explosions and flames, so the same safety precautions with electrical wiring should be followed in these types of buildings as in wood ones. Most types of seeds should be safely stored in structures with these broad features from harvest until the following planting season, except for regions with exceptionally high temperatures and relative humidity. To preserve seed quality during storage in such regions, further precautions—namely, temperature and humidity control—are needed.

Types of Storage Structures

1. **Farm Storage:** During the harvest season, farms typically store their produce for a few days or weeks. On farms, seeds are rarely kept longer than the time between harvest and the following planting season. They could be kept on the ground, in sacks, or in bins. Ground storage is typically utilized in dire situations and is only temporary. Granaries, barns, and wood or metal bins can all be used to store bulk loads. Because metal buildings build up more heat and lose moisture more slowly than wood buildings, seeds, unless they are extremely dry, keep better in wood structures. High moisture seed storage can drastically diminish viability due to heat buildup in metal buildings. Reduced heat buildup can be achieved by ventilation, white paint on the building's exterior, or both. If there is room, seeds in bags can be kept in any farm structure. This method often results in significant losses due to damage from rodents and insects. Seed that is bagged with a lot of moisture in it will heat up and quickly lose all its value. (Barre, 1954; Wheeler and Hill, 1957).

2. **Country Elevator Storage:** A lot of rural elevators store seeds throughout and right after harvest. Storage can be done in bulk, in wood or metal bins, in concrete silos, or in bags or tote boxes inside of buildings made of wood, brick, concrete, or metal. Storage is often done for a brief period. Before storing seeds, country elevators usually feature a fanning

mill to eliminate debris. Additionally, fanning lowers the moisture content of seeds. In addition to possibly having drying equipment, country elevators typically contain capabilities for controlling insects and rodents (Barre, 1954).

3. **Seed Processor Storage:** Collectively, seed processors hold a significant portion of the seeds stored for potential usage in the US. Their storage facilities come in a wide range of sizes and designs, from modest metal or wood structures to enormous multi-story brick, concrete, or wood warehouses. Certain processors, particularly those handling expensive commodities, employ well-insulated rooms with controlled temperature and relative humidity to ensure optimal viability protection. Both processed and raw seeds are kept in storage by seed processors. You can store unclean seeds in bulk bins, pallet boxes, or bags. For a given kind of seeds, each container offers benefits and drawbacks. Typically, unclean seeds should be kept in a location that is clearly marked off from the cleansed seeds' storage space. Seeds that have undergone complete processing are often kept in the shipping container. While the retailer repackages some processed seeds, the processor stores others in retail packets. Seed processors' storage facilities typically meet the previously mentioned basic standards and can also accommodate specific needs of individual seed varieties and provide protection from adverse weather.

4. **Retail Storage:** Seed dealers are required to retain their inventory for variable periods of time, typically within the typical weather conditions of their specific locations. Although retail storage is rarely perfect, seeds in cloth, paper, or thin plastic containers typically have a long enough shelf life to allow marketing with little loss of viability, except for extremely hot and humid climates. Seeds have a short shelf life in hot, humid climates unless they are safely dried to 5 to 8 percent moisture content and placed in moisture-barrier containers. There is little to no effort made at the retail level to offer safe storage.

5. **Research Storage:** For their study, scientists need to keep viable seed stocks in varied numbers, which requires storing. While long-term seed storage is best served by refrigerated, dehumidified storage, research scientists may not always have access to these conditions. Although they usually employ the greatest facilities available, they regularly must work in less than the ideal settings. Many different types of seeds can be kept for several years under normal air conditions in some geographic areas without losing any of their viability. Controlled air storage conditions are necessary in tropical and subtropical regions to maintain research seed

stocks for longer than a few days. Sealed storage is the sole other option. Seeds must be dried to a moisture content of 5 to 8 percent before being sealed in moisture-proof containers for safe, sealed storage for up to 3 to 5 years at room temperature. Before packaging, the moisture content needs to be lowered to 2.5–5% for extended storage.

6. **Germ Plasm Storage:** To reduce regrowing of seed stocks, germplasm preservation centres like the Japanese National Seed Storage Laboratory (Ito, 1972), the U.S. National Seed Storage Laboratory (James, 1972), the U.S. Regional Plant Introduction Stations, and other similar facilities around the world need to have the best storage conditions. Three floors make up the US National Seed Storage Laboratory. The first floor has the biochemistry lab, growth chamber, machine room, control room, custodian's supply room, workshop, garage, and supply storage room. The administrative offices are located on the second floor. The third floor is occupied by the germination laboratory and seed storage rooms. A shared hallway leads to the seed storage rooms, which can hold up to 180,000 pint cans of samples in total. Using specially designed containers can enhance the capacity of the seed sample by a factor of ten. The storage environment is kept at 4 °C on average, with a relative humidity of roughly 35%. Most seed types could be stored in this combination for a fair amount of time. If needed, three of the rooms may be kept at a temperature as low as —12 °C. Smaller rooms not used for routine germplasm storage have a range of relative humidities and temperatures available for research reasons.

Constructing Controlled Atmosphere Seed Storage Facilities

Careful management of the storage area's relative humidity and temperature is necessary for the safe storage of seeds. They can only be managed in specifically designed spaces, such as buildings or rooms. The walls, ceiling, and floor of a seed storage room must have adequate heat insulation and a moisture vapor seal to function as efficient barriers against external sources of heat and moisture. A layer of hot asphalt is often used for installing floor insulation because it creates a good vapor seal. The type of material (Fiber glass, spray-on foam, Styrofoam, cork), the temperature to be maintained, and the amount of insulation needed determine how much of each. Keeping insulating materials dry is essential for optimal performance. Moisture protection needs to be added outside the insulation if the material does not naturally possess a dryness feature. To reduce heat and moisture penetration via the joints, board-type insulation should be put in two or more layers with the joints lapped or spaced apart. The corner flashing vapor seal material has an accordion fold to address the issue of building movement with temperature changes. Many types of ceiling insulation are available. Two coatings of cement plaster, no less than half an inch in thickness, are often used to finish walls and ceilings. Galvanized metal is used to reinforce finish coats in areas of the wall that are susceptible to shock. Bumpers made of concrete, metal, or wood are put on walls where trucks can unintentionally run into them. The doors of cold storage rooms must be tightly sealed and insulated, and they cannot have any windows. Roller hanging doors could be preferable to swinging doors for wide apertures. Not only do roller doors fit more tightly, but they also have electrical operation. Using a cool, high-velocity spray of air across the door's face, often from top to bottom, is a relatively recent concept. While it may not be a perfect solution, this does offer some protection against the entry of heat and moisture. Small anterooms and double-door air locks also aid in lowering the amount of heat and moisture that enters cold storage rooms. The main issues with refrigerated dehumidified storage are heat generated by lighting, heat and moisture produced by humans working in the space, and heat and moisture leaking through the walls, roof, floor, and around the doors. Of course, the best way to minimize these is to build the storage room or warehouse with proper precautionary measures built in. Building multiple controlled temperature rooms is typically preferable to building one sizable warehouse. An important way to reduce annual operating costs is to have multiple separately controlled rooms. When storing tiny amounts of seeds, it is more practical to chill one or two rooms as opposed to the entire warehouse. Most chilled seed storage facilities use forced air that is circulated throughout the space after passing through a cooling coil. A duct system evenly distributes the cold air in a room of considerable size. A qualified refrigeration engineer must make the final decisions on the size and kind of refrigeration system, as well as the structural architecture of the building or room. Installing sufficient insulation, moisture protection, and refrigeration capacity at the outset is significantly more cost-effective and superior than remodelling an inadequate system.

Controlling Temperature

Refrigeration

In general, refrigeration refers to any method of removing heat, therefore the only way to achieve and sustain the low temperature needed for seeds to be stored for an extended period is through refrigeration. Although ventilation can be thought of as refrigeration, it is only appropriate for small temperature decreases. Air circulation must be combined with mechanical cooling to achieve extremely low temperatures. Transferring heat from the body being refrigerated to another body that has a lower temperature than the body being refrigerated is how refrigeration works. Thermal insulation is always required surrounding an area that needs to be refrigerated since heat travels easily from a high temperature area to a lower temperature area.

Heat Load

The pace at which heat needs to be evacuated to create and sustain the required temperature is known as the heat load. The temperature of the area or substance to be chilled, heat leakage through the chamber's walls, floor, and ceiling, heat from lights, motors, and other electrical equipment, and humans working inside the refrigerated chamber are all elements that affect the amount of heat load. The total amount of heat generated by these sources establishes the heat load of a certain refrigeration system.

Refrigeration Agent

The body employed to absorb heat is the refrigeration agent or refrigerant. Cooling systems are classified as either sensible or latent, according to the effect the absorbed heat has on the refrigerant. The cooling process is said to be sensible when the absorbed heat increases the temperature of the refrigerant and latent when the physical state of the refrigerant is changed. With either process the temperature of the refrigerant must always be lower than that of the area or material being refrigerated. Either a sensible cooling system or a latent cooling system can provide continuous refrigeration. Recirculated refrigerant is chilled by a sensible cooling system. Either a liquid or a solid refrigerant, such as ice or solid carbon dioxide (dry ice), can be used for latent cooling. It is not advised to use dry ice or ice as refrigerants in seed storage rooms.

Liquid Refrigerants

The foundation of mechanical refrigeration systems is the vaporization of liquids, which allows them to absorb vast amounts of heat. Liquid vaporizers offer an easily regulated cooling system. These systems can be started and stopped at will, and their cooling rate may be controlled within very specific bounds. By adjusting the pressure at which the liquid vaporizes, one can change the temperature at which the liquid vaporizes. A closed system allows the vapor to easily condense back into a liquid, allowing for repeated usage and a steady supply of liquid for vaporization. The liquid refrigerant that is chosen should be the one that best suits the unique requirements of each storage facility, as there isn't a single liquid refrigerant that works well in all applications and operating environments. Dichlorodifluoromethane (CC12F2) is the fluid that is closest to the ideal general-purpose refrigerant out of all the fluids that are currently utilized as refrigerants. It belongs to a class of refrigerants that were first sold under the trade name "Freon," but are currently produced under several different proprietary brands. Refrigerant-12, or R-12, is the new trade name for this chemical to avoid confusion. Because refrigerant-12 has a saturation temperature of -29.8°C, it can only be kept in strong steel cylinders under pressure when stored as a liquid at normal temperatures. It is not practical to refrigerate a seed storage room by letting liquid R-12 evaporate in a container that is vented to the outside, even though this works well for cooling an insulated chamber. An integral part of any mechanical refrigerating system is the evaporator, a container in which the refrigerant vaporizes during refrigeration.

Typical Mechanical Refrigeration System

The following components are necessary for a typical mechanical refrigeration or vapor-compression system

(1) An evaporator to provide a heat transfer surface where heat is transferred from the space being refrigerated into the vaporizing refrigerant.

(2) A suction line to transfer the refrigerant vapor from the evaporator to the compressor;

(3) A compressor to heat and compress the vapour.

(4) A hot gas or discharge line to transfer the high-temperature, high-pressure vapor from the compressor to a condenser.

(5) A condenser to provide a heat transfer surface where heat is transferred from the hot gas to the condensing medium; The liquid refrigerant is transported from the receiving tank to the refrigerant metering device *via* a liquid line, which is installed in step seven. The refrigerant metering device regulates the liquid flow to the evaporator. There is a lower pressure side and a higher-pressure side in a conventional vapor-compression system. The lower pressure side of the system is made up of the refrigerant metering device, evaporator and suction line; the higher-pressure side is made up of the compressor, discharge line, condenser, receiving tank and liquid line.

Condensing Units

A condensing unit can have a water- or air-cooling system. Small horse power condensing units usually include motor compressor assemblies that are hermetically sealed. Water cooling is typically used for large condensing units.

System Capacity

The rate at which a refrigeration system extracts heat from the chilled area is known as its capacity. The common way to express capacity is in terms of heat extracted per hour or its equivalent in terms of ice melting. Stated differently, a mechanical refrigeration system is said to have a capacity of 1 ton if it can chill at a rate equal to the melting of 1 ton of ice in a 24-hour period. The weight of refrigerant circulated per unit of time and the refrigerating impact of each pound circulated determine a mechanical refrigeration system's capacity.

Compressor Capacity

A compressor's capacity needs to allow for the extraction of vapor from the evaporator at the same rate as its production. The buildup of surplus vapor in the evaporator will raise the pressure, which will raise the refrigerant's boiling temperature, if the vapor is created more quickly than the compressor can remove it. The pressure inside the evaporator will drop and the refrigerant's boiling temperature will drop if the compressor's capacity causes the vapor to be extracted from the evaporator too quickly. The refrigeration systems will not operate correctly in any scenario. The rate of vaporization and the pace at which the vapor condenses back into a liquid must be balanced in any effective refrigeration system. A system with this level of balance will operate correctly and provide excellent refrigeration.

Controlling Humidity

The main gases that make up air include nitrogen, oxygen, carbon dioxide, water vapor, and trace amounts of rare gases. Every gas in the mixture, including water vapor, exerts a different partial pressure as if the other gases weren't there. The mixture's total pressure is equal to the sum of these partial pressures. The temperature and mixture pressure are the only variables that affect the amount of water vapor that can be contained in the air mixture. Since moisture goes from a high-pressure location to a lower pressure area, understanding humidity in terms of partial pressure facilitates understanding of moisture transportation. As a result, moisture can travel against airflow. Although it may slow its penetration, pressurizing a space won't stop moisture from entering. Plotting the intersection of the dry-bulb and wet-bulb temperature measurements on a psychrometric chart is the standard method for measuring relative humidity. A specific relative humidity will be correlated with the junction location. Elevating the temperature of the air can alter the relative humidity but not the absolute humidity.

Moisture Movement Between Air and Materials

While materials like steel and glass retain moisture on their surface, structural elements like wood and cement have moisture throughout. The difference in moisture vapor pressure between such materials and air controls the pace of moisture vapor transfer between them. Moisture cannot transfer from one to the other if their moisture vapor pressures are equivalent. Moisture will seep from moist seeds into a dry atmosphere when they are placed in it. The air will quickly get saturated with the moisture that the seeds release since it is unable to keep nearly all of the moisture that the seeds contain. Unless fresh, dry air is supplied, the seeds will cease to dry. Moisture in building materials must be eliminated, in addition to the moisture present in the seeds. The drying system's function is to eliminate any moisture that enters the controlled environment room through door openings, leaks through seams and cracks, and barrier material penetration once the room and the seeds have reached moisture equilibrium with the specified relative humidity. Once seeds have achieved moisture equilibrium with the maintained relative humidity, the storage compartment is kept at a specific relative humidity, which in turn prevents any change in the moisture content of seeds.

Refrigeration-Type Humidity Control Systems

Warm, humid air is drawn over a metal coil including fins that are spaced widely enough to allow for partial icing and adequate airflow by the refrigeration-style dehumidifier. A timeclock-controlled electrical heater, hot gas, or water defrosting at regular intervals can all be used to eliminate the frost. A refrigeration-style dehumidification system needs to cool the air below the target temperature and reheat it to the target temperature in order to function well at low temperatures. Electric defrosters, reheat coils, and refrigeration coils integrated inside the unit are options for air-handling units. Because the constructed device provides the correct wattage, it can be preferable to a fragmented installation. Connecting the reheat coils to a humidistat *via* a relay is the most straightforward way to control the reheat in a space. While the refrigeration-re-heat system is a useful tool for controlling temperature and relative humidity, it is not the only option and might not always be the best. The refrigeration-reheat system becomes extremely costly to operate when extremely low relative humidity is required.

Desiccant-Type Humidity Control Systems

Refrigeration combined with liquid or solid desiccants, used in dehumidifiers, can often lower the cost of maintaining very low relative humidities. After absorbing moisture vapor from the airstream, these desiccants release it outside the space. For small systems, desiccant dehumidifiers typically employ dry chemicals; for systems with very high volumes of air, they use salt solutions. Systems that use dry desiccant are nearly always utilized in seed storage facilities. One or two beds of activated alumina or granulated silica gel, which are highly capable of absorbing water vapor, are incorporated into the dehumidifier. For instance, at 100% relative humidity, silica gel may absorb up to 40% of its weight in water vapor; with lower relative humidity, this capacity decreases proportionately. One bed at a time, the two-bed method that is most utilized in seed storage facilities, allows for air circulation. Whereas the other bed is absorbing wet, the recharged bed is dry. Typically, a timeclock is used to program the transition between beds for optimal efficiency. Rotating disc and cylinder dehumidifiers are two recent innovations that are finding extensive use. One or more beds in the rotating dehumidifiers are separated into two airstreams by sealing strips. A portion of each bed is recharging while the other is absorbing water vapor from the airstream as the bed, or beds spin slowly. Except for the rotary systems' apparent higher drying, the final product is essentially the same as for the two separate bed systems. A combined refrigeration-desiccant dehumidifier system would likely offer the most dependable temperature and relative humidity control at the lowest cost for seed storage facilities that require both a low temperature and a low relative humidity. The desiccant material in desiccant systems needs to be replaced every three to five years.

Common-storage Practices

Airflow whether they are kept in bags or in bulk, moist seeds tend to heat up unless the heat is quickly released as it is produced. Through ventilation, extra moisture and heat can be released. Heat is captured and transported away by a constant flow of air passing through a warehouse filled with bags of seeds or a bin of loose seeds. Keeping the temperature from rising too high is crucial for healthy germination. Seed quality will soon pay for a well-designed ventilation system. Many processed seeds are worked with in bags and kept in stacks. Each genus, species, and cultivar of seeds is kept in a different stack of bags. Stacks of seed bags should be widely separated, both between the bags and between the stacks, to allow for enough ventilation. Care must be taken when stacking seed bags to avoid slipping and falling. Employees are at risk from falling bags, which can also shatter upon hit. Overly tall stacks frequently result in the bottom bags bursting. It is particularly crucial that pallets be used to properly stack bags that are carried by forklift.

Removing seeds from Controlled Storage

When dry seeds are taken out of cold storage and placed in a warm, humid environment, they will quickly absorb moisture. If precautions are not taken, condensing ambient moisture will cause sacks of cold seeds to become wet. Moisture condensation can be avoided by either heating the seeds in a dry environment or in a room with fast-moving fans. Condensed moisture can be kept away from the seeds by using moisture-barrier containers. Cold bulk seeds can be brought to room temperature by proper aeration without accumulating condensed ambient moisture on their surface. The protective benefits of controlled storage will soon wear off if seeds maintained in a controlled atmosphere are not kept dry after being removed from storage.

Warehouse Cleanliness

Seed quality can be significantly impacted by warehouse cleanliness, or lack thereof. Rats and insects are easier to manage in a clean warehouse since there are fewer of them present. Accidents involving staff and inadvertent mixing are less likely in a clean warehouse. Visitors are also given a better impression of the company.

Factors Influencing seeds Longevity in Storage

1. **Kind / variety of the seed:** The type and variety of seeds have a big impact on how well they store. Some types, like onions, peanuts, soybeans, and others, have a short lifespan by nature. Storability is also influenced by the genetic composition of the lines and variants within a species.

2. **Initial seed quality:** Strong, undamaged seed lots keep their seeds fresher longer than seed lots with degraded seeds. Even quantities of well-germinated seed can and will deteriorate quickly within a few months, depending on the degree of degradation or severity of damage, such as the amount of weathering damage, mechanical injury, or flat, wrinkled, or otherwise damaged seed. This has the significant consequence of indicating that only superior seed ought to be preserved. The seed of average quality might only be kept for the upcoming plating seasons. The inferior seed ought to be rejected without exception.

3. **Moisture content:** Over much of the moisture range, the rate of deterioration increases as the moisture content on seed storability, making the quantity of moisture in the seeds likely the most significant factor determining seed viability during storage. Moisture content and

seed vigour at storage start for cereals with high germination and high temperatures not to exceed 90° F. It is noteworthy that seeds may suffer from severe desiccation if their moisture content falls below 4%. It is essential to dry seeds to safe moisture contents because a seed's life and duration are mostly determined by its moisture content. However, the length of storage, the sort and variety of seed, the type of storage structure, and the type of packaging material used all affect the safe moisture content.

Seed moisture content (%)	Storage life
11-13	½ year
10-12	1 years
9-11	2 years
8-10	4 years

4. **Relative humidity and temperature during storage:** Temperature and relative humidity are by far the most significant variables affecting how long seeds can be stored. When exposed to precise levels of ambient humidity, seeds acquire a moisture content that is distinct and unique. For a certain type of seed at a given relative humidity, this moisture content—known as equilibrium moisture content—tends to grow as temperature drops and degradation advances. Hence, at equilibrium moisture content, there is no net gain or loss in seed moisture content. Instead, the maintenance of seed moisture content during storage is a function of relative humidity and, to a lesser extent, temperature. When seeds are placed in environments with relative humidity levels that differ from the equilibrium of their moisture content, they will experience moisture gain or loss until they reach a state of equilibrium with their new surroundings. The relative humidity within the sealed storage containers is determined by the moisture content of the seeds. In seeds, the process of establishing moisture equilibrium is time dependent. It doesn't happen right away. It takes a while, and how long it takes depends on the type of seed, its starting moisture content, the temperature, and the typical relative humidity. The moisture content of seeds varies when they are stored outside due to variations in relative humidity. On the other hand, moisture content is not much impacted by typical daily variations in relative humidity. Temperature appears not to be a regulating factor, but it does play a significant effect in the life of seeds. The biological activity of seeds, insects, and molds rises with temperature within their typical range. The seeds are more negatively impacted by temperature the more moisture they contain. Therefore, lowering temperature and seed moisture is a good way to keep seed

quality intact while it's being stored. Even though there may be a significant relative humidity increase, low temperatures are particularly helpful in preserving seed quality. Relative humidity in cold storage for seeds shouldn't be higher than 60%. Harrington proposed the following straightforward guidelines, which serve as a helpful gauge for the effects of temperature, relative humidity, and moisture content on seed aging:

a) The storage potential of seed is almost doubled for a 1% drop in moisture content.

b) The amounts of seeds that can be stored almost doubles with a 10 °F drop in temperature.

c) Proper seed storage is accomplished when the temperature in degrees Fahrenheit and the percentage of relative humidity in the storage environment equal one hundred, with the temperature contribution not to exceed fifty degrees Fahrenheit.

5. **Effects of fluctuating environment conditions on viability:** Although there have been a few reports suggesting that unstable conditions are detrimental, there is currently no solid evidence to suggest that variations in temperature or moisture content would be detrimental on their own, except for extremely quick changes in seed moisture content.

6. **Oxygen pressure:** According to recent studies on the effect of a gaseous atmosphere on seed viability, longer periods of viability are typically associated with higher oxygen pressure.

7. **Genetic:** Inheritance of seed longevity is not limited to species. Several cultivars of a species also differ significantly for seed longevity. The seeds of different plant species vary widely in their lifespan under identical or favourable storage conditions.

8. **Preharvest factors:** Temperature, photoperiod, mineral nutrition, rainfall and soil moisture are important factors for seed viability and storage.

9. **Seed maturity:** There is strong relationship between seed maturity and storability or longevity. The healthy, mature, plump seeds generally store better than immature seeds.

10. **Seed dormancy:** In some seeds, dormancy is caused by seed structure, physiology of the embryo, germination inhibitors and combination of these factors. Storability can be affected by dormancy.

11. **Mechanical damage:** During harvesting and threshing, seeds can be damaged. Damaged seed, even small and hidden injuries should not be stored.

12. **Vigour:** The vigour of the seeds at the time of storage is an important factor that affects their storage life. The decline in vigour and viability of seeds is sometimes illustrated by a sigmoid survival curve. The survival curve for dry seeds under favourable environmental conditions can be divided into three distinct parts. The first represents the period when the seed is vigorous and decline in the life functions has preceded .slowly. Eventually, this stage ends at a survival level of 90-75 percent and deterioration proceeds very rapidly. After deterioration has proceeded to a survivable level of 25 to 10 percent, it slows again and continues slowly until all seeds are dead.

13. **Other factors:** In addition to the factors already mentioned, other factors that impact seed storage life include exposure to direct sunlight, the type and frequency of fumigation, the impact of seed treatment, etc.

References

Barre, H. J. 1954. Country Storage of Grain. In Anderson, J. A., and Alcock, A. W., eds., Storage of Cereal Grains and Their Products, pp. 308457. St. Paul, Minn.

Edwin J. 1972. Organization of the United States National Seed Storage Laboratory. In Roberts, E. H., Viability of Seeds, app. 1, pp. 397-404, illus. London.

Harrington JF (1972) Seed storage and longevity. Pages 145–245. In: Kozlowski TT (ed) Seed biology. Insects, and seed collection, storage, testing, and certification. Physiological ecology. A series of monographs, texts, and treatise. Vol III. Academic Press, New York and London

Ito, H. 1972. Organization of the National Seed Storage Laboratory for Genetic Resources in Japan. In Roberts, E. H., Viability of Seeds, app. 2, pp. 405-416, illus. London.

Roberts EH (1973) Predicting the storage life of seeds. Seed Science and Technology, 1:499–514.

Royal Botanic Gardens Kew (2020) Seed Information Database (SID). Version 7.1. Royal Botanic Gardens Kew, Richmond, United Kingdom http:// data.kew.org/sid/ (accessed Mar 2020)

Wheeler, W. A., and Hill, D. D. 1967. Grassland and Seeds. 734 pp., illus. Princeton, N.J.

www.hasannuzaman.web.com

3

Principles and Practices of Seed Grading and its Different Grading Techniques

Meghali Barua

Objectives of Seed Grading

Minimization of heterogeneity to the lowest level and enhancement of the homogeneity to the highest level in the seed lots received by the processing unit.

Principle

Separation of seed lots based on differences in the physical properties of the seeds.

Post harvest activities include threshing/ seed extraction, drying, pre-cleaning and conditioning, cleaning, fine-cleaning or grading, seed treatment, weighing and bagging/ packaging, sealing and labelling, storing, and transportation for marketing. Except for transportation and marketing, all these activities collectively known as seed processing. However, for most of the crop, threshing or seed extraction, basic cleaning even drying is done in the farm itself and in the processing unit, the threshed seed is received.

The objective of the seed processing is to prepare the harvested seed in such a way that seed of highest purity and quality would be available for marketing, also for planting. The cleaning and the grading are the foremost activity of the seed processing. The seed, which is received in the processing unit, contains many undesirable particles, which as such cannot be used for marketing, hence cleaning and grading become necessary.

Seeds received by processing unit, is highly heterogeneous, reasons for which are many.

1. Soil variation in physical, chemical, and biological properties.
2. Soil variation in fertility

3. Inefficacious agronomic practice, e.g., land preparation and levelling, irrigation, fertilizer, or pesticide applications etc.
4. Genetic make-up of the plants, especially in heterogeneous population.
5. Position of the seed (sink) with respect to photosynthetic apparatus (source).
6. Infestation by pests and diseases.

Efficient processing leads to homogeneity in the seed lot, enhances the seed quality in terms of moisture content, purity, germination, and health, moreover it improves the visual appearance thereby increasing the market value; helps in reducing seed rate; helps in uniform seed germination, crop stand and high yield. It also leads to longer shelf life of the seed.

Quality Seed

Impurities that may hamper the quality of pure seeds are

• Inert matter	• Other distinguishable variety seed
• Other crop seed	• Damaged seed
• Common weed seed	• Deteriorated or disease infected seed
• Obnoxious weed seed	• Shriveled or undersized seed • Ill-filled or immature seed etc.

Seeds which are free from all these impurities are regarded as pure, healthy, and quality seeds.

For certification seed must meet following quality standards, which are tested in the seed testing laboratory after completing of processing

- Moisture content (optimum %)
- Physical purity (min. %)
- Germination (min. %)
- Vigour
- Seed health (max. incidence of diseases or insects)
- Genetic or varietal purity (min. %)

Hence, while processing utmost care should be taken to enhance the quality and health of the seed.

Cleaning and Grading

Pre-cleaning removes the impurities which are very large or small or lighter than the pure seed, whereas conditioning prepare the seed for better cleaning, e.g., separating maize seed from cob, removal of awns in wheat, separating tightly held husk from grass seed etc. Additionally, it reduces the seed lot size for subsequent cleaning process. Seeds which have been winnowed or partially cleaned in the farm itself, need not to go through pre-cleaning process.

Cleaning of pre-cleaned seed is done to separate all the undesirable particles which would further improve the physical purity of the seed.

Fine-cleaning or grading of cleaned seeds leads to highest possible level of pure seed. The cleaned seed may not be uniform in size or shape or density etc. or it may contain contaminants which are very similar to the seed in physical properties. The grading separates the seed into different quality fractions depending on the various commercial uses, which is done based on physical differences present among pure seed and the impurities or contaminants. It not only removes under- or over-sized seed but also removes shrivelled, damaged, or cracked or defective seed from the cleaned seed.

It is a proven fact that seed germination and vigour are directly proportionate to the physical traits such as seed size, length, density etc. The grading grouped the cleaned seed into different range of size, shape, density etc., thus upgrading the quality of the seed lot.

The physical traits of the seed that have been exploited by the processing equipment for grading are

Physical traits of the seed	Grading equipment
Seed size (width, thickness, small to bold)	: Air screen cleaner cum grader
Length	: Disc or indented cylinder separator
Shape (round/ flat/ oblong seeds etc.)	: Spiral separator or draper separator
Seed mass (weight, density-light to dense)	: Specific gravity separator
Surface texture (smooth or rough)	: Seed-dodder mill, roll mill
Colour (mature or immature, light, or dark)	: Electronic colour separator
Electrical conductivity (low or high)	: Electronic separator
Filled or ill filled seed	: Liquid floatation method
Surface texture, stickiness	: Magnetic separator
Aerodynamic properties (critical velocity, drag coefficient and Sailing velocity)	: Through air stream

Based on variation in physical traits, various equipment can be used for grading.

Grading Equipments

The most common equipment used for grading are indented cylinder, gravity separator which are placed after air screen cleaner, others are disc separator, spiral separator or draper separator, seed-dodder mill, electronic colour separator, electronic separator, magnetic separator etc. Digital techniques can also be used for real-time analysis.

Air Screen Cleaner cum Grader

The air screen cleaner cum grader separates the undesirable particles from the seeds based on differences in seed size, weight, and shape. It uses multiple screens of various mesh size and, aspiration for cleaning and grading. The screen hole may be round, oblong, triangular depending on the seed shape and impurities (Appendix-I). There are three cleaning constituents-

i. **Aspiration (air flow)**: With the help of air blow the lighter impurities are removed from the pure seed.

ii. **Scalping**: Desired seeds are allowed to pass through the screen of specific mesh size and larger impurities remains on the screen which is discharged through outlet.

iii. **Grading**: Desired seeds remain on the screen, but smaller impurities pass through the screen which is discharged through the outlet.

Indented Cylinder Separator

The indented cylinder separator separates the undesirable particles from the seeds based on differences in length. Based on the requirement the indented cylinders can be arranged as single unit; or as multiple units in a series or parallelly; or in such a way that one cylinder will remove longer impurities, and another will remove shorter ones. This helps in removal of broken seeds also.

When seeds are fed into the hopper, it moves through the rotating cylinder, due to centrifugal force the materials smaller than the pure seed will be carried up in pockets i.e., indents which are present in the inside surface of the cylinder and whose size are selected based on the material that must be removed. These small sized particles which may include broken seed then collected in a trough and discarded out through a conveyor. Seeds larger than the pockets remains in the cylinder moves out by gravity and are collected separately. While operating, the feeding rate should be uniform and optimum for best efficiency.

To test the efficiency of the separator, the seed sample can be taken from the discharge outlet for the pure seed.

Specific Gravity Separator

The specific gravity separator separates the undesirable particles from the seeds based on differences in size and weight or density, i.e., seeds of same size but with density variation and of similar density but with size variation. It is based on floatation principle which separates the cleaned seed into different quality fractions based on their specific weight. Specific gravity separator is crucial operation for high value crops (e.g., vegetable seeds, pulses etc.), though crops such as cereals need not to go through it once the seeds cleaned and graded by indented cylinder since it would not be economically remunerative.

The equipment consists of inclined deck made of square wire mesh mounted on a frame. The deck, which may be rectangular or triangular, is fitted in such a way that it can oscillate. The deck can remove both lighter and heavier particles, but the rectangular deck can remove heavier and triangular lighter particles more efficiently. There is a fan below the deck to blow the air. The deck inclination, oscillation speed, air flow all can be adjusted to match the seed type that has to be upgraded.

The seed is fed uniformly on to the deck surface. The air blown below the deck causes lighter particles to float in upward direction whereas heavier seeds remain on the deck thus forming different layer based on the density. When deck oscillates, heavier seed which are on the deck move uphill and discharged at the upper end, and the lighter seeds due to gravity move down at the lower end and discharged. In the end, seeds can be collected from the different discharging outlet in a series having from light to heavy weight.

Spiral Separator

The spiral separator separates the undesirable particles from the seeds based on differences in shape i.e., it separates the seed based on their ability to roll. Initially it was used for soybean seeds, but now used for any seed which are spherical e.g., brassica seeds.

The seed is fed at the top of the column with a spiral slide. The more spherical the seed, the faster it will roll down the column than the non-spherical particles. The speed of the spherical seed increases while going down the column due to which at one point they roll over the edge where they are collected. Other seeds continue to roll down through the column but slowly and get collected in separate trough.

Velvet roller/ Seed-dodder Mill

The velvet roller separates the undesirable particles from the seeds based on differences in texture (smooth or rough) i.e., it uses the property of friction, higher the roughness, more will be the friction. It can separate the seeds and contaminants having same size and density but show difference in roughness. It is mainly used for forage crops e.g., in clover, seed maturity is heterogeneous due to variation in flower patterns, hence some seeds become wrinkled after drying. It can even separate dodder seed (*Cascuta* spp.), a weed, from the clover or alfalfa seeds.

The equipment has two attached but parallel cylinders which is covered by a cloth made of natural fibre. The cylinders are slightly inclined and rotate in the opposite directions at 0.5-1 revolutions/ second.

The seed when placed between the two rotating cylinders, it moves down the slope over a belt made of either velvet cloth or rubber having variable inclination providing variable friction to the moving seeds and contaminants, this rotational movement lift the seed with rough surface, which are collected in different troughs based on various degree of roughness.

Disc Separator

Like indented cylinder separator, disc separator also separates the undesirable particles from the seeds based on differences in length. It is more precise than the indented cylinder but with very low output.

The disc separator consists of many discs arranged vertically on a central shaft at 5-10 cm apart. These discs are fitted with pockets where shorter seed get collected upon rotation of the disc.

Electromagnetic Separator

The electromagnetic separator separates the undesirable particles from the seeds based on differences in the seed surface texture i.e., even, or uneven surface. It is a sophisticated process hence it is used for very expensive seed only.

When seed is fed into the hopper, there is a mechanism that adds iron powder over the seed. Water can also be added for better adherence. The iron powder gets attached to the seed having uneven surface. The seed is then poured on to a rotating electrically magnetized metallic drum. Iron coated seed get stick to the drum for longer duration compared to the seed with an even surface due to absence of iron coating. Thus, seeds are collected in different containers.

Colour Sorting

The colour sorting separates the undesirable particles from the seeds based on differences in the seed colour which may be of same size and density. The differences in the colour may be due to differences in maturity of the seed, since immature seed would contain chlorophyll; the unhealthy seed would be dull than the healthy seed, may be due to seed aging in storage etc. Off coloured seed can be removed both manually and mechanically.

The colour sorting also known as optical sorting is a sophisticated process and based on light reflection. Seeds that are fed into the hopper are steered into a thin line on the conveyer, which then passes through a dark chamber, fitted with a sensor. In the dark chamber each seed is exposed to a specific light, any seed or impurity showing deviation in colour will be detected by the sensor and a high-pressure air will blow away that impurity from the path of the flow. Thus, both pure seed and impurities are collected separately which are discharged from the separate outlet. The new version has digital image analyser which are more efficient. Like, electromagnetic separator, this equipment is also very expensive with low output, hence used for highly expensive seed only.

Liquid Floatation Method

The liquid floatation method separates the filled seed and ill-filled seed based on differences in the density. It is the simplest method of upgrading which can be followed in the farm itself, especially used for paddy seed.

Seeds are soaked in a liquid of specific gravity, due to which filled seed will sink to the bottom whereas ill-filled seed or other impurities will float, which then can be separated easily.

Polishers

The polisher improves the lustre of the seeds e.g., beans, peas, popcorn etc., either by using polishing agent (sawdust, bran etc) or by mechanical brushing or rubbing the seed. In the first case polishing agent and seed is mixed in a chamber which is then agitated by conveyer, after polishing seed is passed through a screen to separate the polishing agent.

Picker Belts

Those impurities that cannot be removed by machine but by hand, in that case, seeds are placed in a moving belt, where handlers observe the seed and removes the impurities manually. e.g., damaged corn ear, ground nutshell etc.

Forages		
Berseem: Diploid	2.00 r	1.00 s
Forage sorghum	4.00 r, 4.75 r	2.10 s
Guineagrass	2.10 r	2.40 x 0.65 m
Lucerne	2.50 r	0.70 x, 0.70 x 0.70 m
Oats	7.50 r	2.00 s
Vegetable crops		
Cucurbits: Ashgourd	9.50 r	6.40 r
Bittergourd, Bottlegourd, Pumpkin	11.00 r	6.50 r
Cucumber	8.00 r	2.00 r, 2.50 r
Indian squash	9.50 r	6.40 r
Longmelon	5.00 s	1.00 r
Muskmelon	5.00 s	1.00 r
Snakegourd, Ridgegourd, Spongegourd	9.50 r	6.40 r
Snapmelon	5.00 s	1.00 r
Summer squash	8.00 r	2.00 r
Watermelon	6.00 r	1.80 s
Fruit Vegetables: Brinjal, tomato	4.00 r	0.80 s, 2.10 r
Capsicum, Chilli	4.00 r	0.80 s, 2.10 r
Okra (Bhindi)	6.00 r	4.30 r
Rat-tail radish	4.50 r	2.00 r
Leafy Vegetables: Asparagus	6.00 r	2.40 r
Celery	1.80 r	0.40 s, 0.64 x 0.64 m
Fenugreek: Large/ Medium	3.25 r	1.20 s
Small	2.10 r	0.69 x 0.69 m
Lettuce	2.30 r	0.80 r
Parsley	2.75 r	0.75 s
Spinach beet	5.50 r	1.80 s, 1.85 s, 2.25 r
Spinach: Round seeded	5.00 r	2.75 r
Spinach: Sharp seeded	8.00 r	2.5 r
Coriander	4.25 r	2.5 s
Cole Crops: Cabbage, Chinese cabbage	2.75 r	0.90 s
Cauliflower, Broccoli, Knol-kohl	2.75 r	1.10 s
Bulbs crops: Onion	3.80 r	2.00 r
Root Crops: Carrot	2.30 r	1.00 r
Garden beet	9.00 r	3.00 r
Radish	4.50 r	2.00 r
Turnip	1.80 r	1.20 r
Sugar beet: Monogerm	9.00 r	3.00 r
Sugar beet: Multigerm	9.00 r	2.50 s

(***Source**: Indian Minimum Seed Certification Standards, 2013*)

m= wire mesh sieves; r= round perforated screen; s= slotted/ oblong perforated screen

References

Agrawal RL (2013). Seed Production of oil crops. In book: Seed Technology, Publ.: Oxford & IBH Publishing Co. Pvt. Ltd. pp: 159-182.

Bosoi ES, Vemiaev OY, Smimov BG and Sultan-Shakh. (1990). Theory, Construction and Calculations of Agricultural Machines: Volume-Il. Oxonian Press Pvt. Ltd., New Delhi.

Brini M (2023). Digital Agriculture for seed grading. Digital Agriculture: Grain Quality Analysis.https://www.researchgate.net/publication/370466266_Digital_Agriculture_for_seed_grading

FAO (2018). SEEDS TOOLKIT. Module 2: Seed processing: principles, equipment and practice, Publ.: The Food and Agriculture Organization of the United Nations and Africa Seeds, Rome. pp: 41-48.

Indian Minimum Seed Certification Standards (2013). The Central Seed Certification Board, Department of Agriculture & Co-operation, Ministry of Agriculture, Government of India, New Delhi.

McCabe WL, Smith JC and Harriott P. (1993). Unit operations of Chemical Engineering. McGRAW-HllLInc., New York.

Phukan A, Mahanta M and Barua M (2024). Artificial Intelligence for seed quality control. Book: Smart Agriculture-Digital Era in Farming. Publ. Elite Publishing House. pp: 216-225.

Potty YH and Mulky MJ. Food Processing. Oxford & IBH Publishing Co. Private Limited, New Delhi.

Sahay KM and Singh, KK (1994). Unit operations of Agricultural Processing. Vikas Publ. House Pvt. Ltd., New Delhi.

4

Principles of Seed Storage and Factors Influencing the Life Span of Seeds During Storage

Sharmila Dutta Deka and Priyanka Sharma

Objective of Seed Storage

The primary goal of seed programs, which include storage, is to maintain the high standards of the seed from harvest until the crop is sown in the following or future seasons. Additionally, seeds are kept in storage for longer periods of time for a variety of reasons, including protection against crop failures during emergencies or natural disasters, maintaining active genetic stocks for breeding, germplasm preservation for long-term use, stock preservation for the domestic and international seed trade in response to market demands, and parental line seed preservation for the transnational seed industry during one or more growing seasons. The seeds of most major agricultural species can be classified as either orthodox or desiccation tolerant. A drop in storage temperature and relative humidity, commonly referred to as seed moisture content, shortens their shelf life. Seeds deteriorate even in dry storage because of alterations caused by pests and pathogens as well as physiological deterioration. This is accurate, even though the amount of time seeds takes to germinate and stay viable varies according to the makeup of the different plant species. A good seed program handles the seed with care, regulates the store's temperature and relative humidity (or seed moisture in the case of hermetically sealed containers), and adheres to good sanitation practices to preserve the high planting value of the seed in terms of purity, germination, vigour, and seed health.

Concept of Seed Storage

When a seed reaches "physiological maturity," its germination and vigour reach their maximum point in development. The seed crop is permitted to dry properly at this phase due to the high seed moisture content. According to

Verdier *et al.* (2020), this stage, also known as "field maturity" or "harvest maturity," occurs when the seed's moisture content reaches an acceptable level (less than 15-20%, depending on the species) without being too dry to break. However, research conducted in the northwest of India on several winter crops (rabi season) has revealed that seed moisture levels during a dry season might decrease to as low as 5-6% at harvest. Conversely, harvested seed from crops grown during the wet season sometimes requires an extended drying period prior to processing. Most seeds used in horticulture and agriculture have a typical behaviour: they get longer as the seed moisture content drops and may be stored at low temperatures for longer periods of time. Since seeds are living organisms, harvesting and subsequent storage cause their viability and vigor to gradually decline—at first very slowly—even on the mother plant when the seeds reach physiological maturity. The seed eventually loses its capacity to sprout and yield a robust seedling (Ellis, 2019).

Purpose of Seed Storage

The seeds from the healthiest plants in the crop are removed, cleaned, dried completely, and kept in a cool, dry place in preparation for planting the crops the next season. Seed technology has advanced significantly in tandem with the specialized science of agriculture. Ensuring that seed with high germination and vigour is accessible for the upcoming planting season is the primary objective of storage, regardless of whether it is farm saved or held by organized seed producers. More than 75% of the yield is used for this purpose (based on Indian seed system). Moreover, government organizations or commercial producers reserve a portion of the overall seed stock, especially the food security crops, to meet seed demands in the event of a calamity. In response to expected demand in other regions, those involved in both domestic and international seed trades also maintain certain stocks on hand for future use. Additionally, to lower production costs, seed manufacturers typically keep the hybrids' parental lines' seeds in medium-term storage for two seasons or longer. Breeding lines and small amounts of germplasm seeds are kept in long-term storage. Planning for appropriate and economical storage management is necessary to achieve these goals.

Factors Affecting Seed Storage

Seed Factors

These encompass all internal elements that impact the seed's physiological status both during and after harvest. These consist of:

- The seed's genetic composition.
- Conditions under which the seed crop grows, any biotic or abiotic stress, or dietary deficiencies, particularly while the seed is developing.
- Seed quality at harvest, encompassing maturity, size, and physical damage.
- The moisture content of seeds throughout storage and processing.
- Weed seeds, internal insect infestation, and pathogen burden.

Storage Factors

These are basically the outside variables that have a big impact on how long a species' seeds store.

- Relative humidity, or moisture in seeds.
- The outside temperature.
- The atmosphere is gaseous.
- Materials for packaging.
- The seed retailers' layout and cleanliness.

In addition to the great genetic diversity, it is vital to monitor seed qualities that affect storability during the mother crop's raising. During the seed maturation process, the crop must be produced in an agro-climatically favourable environment with moderate temperatures and dry weather. It must also have an adequate supply of nutrients in the soil, be free of diseases and pests that propagate through seeds, and not be subjected to biotic or abiotic stressors throughout the overall growth of the crop or during the maturity of the seed. Finally, when the seed crop reaches the appropriate maturity stage, it must be harvested. Hard seeds, reduced seed viability, and vigor in soybean are caused by both biotic stress from Phomopsis during the seed-fill and pre-harvest stages and abiotic stress from high soil temperatures and moisture (TeKrony *et al.* 1996). The percentage of germination for a particular seed batch will decrease due to the unavoidable physiological deterioration of seeds during storage. Moreover, pests and viruses cause harm that lowers the seed's planting value. If a seed lot exhibits a high level of pest infestation or a germination rate significantly below the suggested threshold for a particular crop species, it is considered unsatisfactory. These seeds are thrown away as trash because, if they have been treated with fungicides, they are not even fit for use as food or feed. As a result, there is a significant waste of resources due to germination loss and damage from pests and diseases. Inadequate storage conditions lead to seeds losing quality, which might lead to a 25% annual loss of seed inventory, according to McDonald and Nelson (1986). This loss may be significantly greater in tropical and sub-tropical regions (Champ 1985). The purpose of storage is achieved by attempting to slow down the rate of deterioration when necessary, using the best packaging and storage procedures. Degradation cannot be totally controlled or avoided.

Management of Storage Factors

Temperature, relative humidity (RH), and the gaseous composition of the storage environment are the three most significant variables that influence seed storability directly, indirectly, and through interactions between biotic and abiotic factors. These variables are listed in order of importance. Pérez-García *et al.* (2007) reported that ultra-dry (0.3–3.0%) and low temperature (-5.0–5.0 °C) conditions preserved the germinability of orthodox seeds for lengthy periods of time; these are usually recommended for the long-term storage of modest amounts of seed, such as in gene banks. The gaseous composition of the storage space affects seed storage as well, especially in hermetic environments (Groot *et al.*, 2015). Water potentials of dry seeds typically range from -350 to -50 MPa. Due to their high hygroscopicity, seeds take in moisture from the damp air and hold it until they reach a state of equilibrium. However, in a particular climate, the pattern of moisture absorption is influenced by the chemical makeup of seeds. According to Thomson (1979), at 50–60% relative humidity, the moisture content (MC) of most cereal seeds and pulses achieved safe values of 11–13%. At all temperatures, there were differences in the water absorption and desorption curves; the former had a steeper slope than the latter. Because of this, unconditioned shops have a lot of RH fluctuations, which causes the seed to absorb and desorb moisture vapor before it reaches an ERH. All these factors need to be considered while making plans for storing seeds. Ground breaking study on seeds of different crop species stored at different ranges of MC/RH and temperature was carried out by EH Roberts and his team at the University of Reading, UK. As a result, viability equations were created to forecast whether seeds of a particular species will germinate at a particular temperature and moisture/humidity level (Roberts 1973; Ellis *et al.* 1989). To forecast the storage behavior of particular species more accurately, the underlying equations were further adjusted. Nevertheless, there are serious concerns regarding the validity of these equations because they rely on species-specific moisture constants and universal temperature constants between -13°C and 90°C (Pritchard and Dickie 2003). Important factors that they neglect to consider include genotypic variability within species, the glassy state of the cytoplasm at MC of 5% and below, and gaseous compositions in sealed storage. It does, however, emphasize the significance of humidity and temperature (seed moisture) in seed storage and provides broad markers for the viability of storing seeds of different species. Even while they both matter and influence how long seeds endure, moisture—especially in warm climates—is more significant and manageable in terms of energy than cooling. Omobowale *et al.* (2016) stated that temperature is not as important in preserving any kind of seed or grain as excessive humidity is. This means that without requiring the energy or infrastructure modifications needed for refrigerated storage, even small drops in ERH can have the same effect as a drop in temperature (Dadlani *et al.* 2016). The MC attained by a seed lot equilibrated at a given relative humidity, or ERH, is what is known as equilibrium relative humidity. According to Bradford *et al.* (2018), bacteria can proliferate up to 90% ERH. They also proposed a connection between different microflora and insect pests that thrive on seeds with a decreasing moisture content. Although active respiration of seeds stops below about 95% ERH. However, due to their ability to both manufacture water through metabolic pathways from their feed and prevent water loss, storage insects may remain active at ERH below 65% while fungi cannot. However, they are not able to live at ERH concentrations below 35% (Roberts, 1972). On the other hand, seeds need to be dried to 6 ± 1% seed moisture, packed in moisture-tight containers, and stored at room temperature or 25 °C to reach ERH values. Important plant species with limited storage lifetimes can be preserved for medium- to long-term storage under low moisture conditions (Ellis and Hong 2007). The significance of temperature and seed moisture (as well as relative humidity) was emphasized by Harrington (1972), who also put forth two basic guidelines: a seed's storage life doubles for every 1% drop in seed moisture and every 5°C drop in temperature. These two variables' effects are independent of each other. Harrington (1973) stated that seeds of most crop species can be safely stored for one to three years if the temperature and relative humidity are maintained between 100 and 110° Fahrenheit. However, if the temperature does not account for more than half of the RH and temperature (in °F), seeds can be stored at these conditions without harm (Justice and Bass 1978). It is possible to manage storage elements to varying degrees using artificial or natural ways. It is advised to arrange the seed material in sacks and bags on wooden pallets, making sure to maintain a safe distance from the ceiling and walls. Depending on the kind of seed and its intended use, several storage configurations are considered in commercial shops to maintain seed quality.

Management of Moisture

To do this, there are two methods: (a) Maintain the seed in balance by keeping it dry and the equilibrium relative humidity (ERH) low, allowing the seed to maintain a predetermined moisture content; or (b) dry the seed, which entails bringing its moisture content down to a predefined low point before placing it in a moisture-tight, hermetically sealed container. According to studies, food products fare better in environments with lower equilibrium relative humidity (ERH) because desiccation inhibits the metabolic processes of bacteria, fungi, and insects that lead to rotting (Crowe *et al.* 1992). However, at a given ERH, higher oil content seeds (or grains) will have a smaller MC than lower oil content seeds. This is because the hydrophobic oil bodies in the cells reject water, resulting in a final output with a decreased total water content.

Likewise, this applies to seeds. An "isotherm" is the temperature at which the connection between MC and ERH in seeds of a certain species stays constant. According to Bradford *et al.* (2018), when it comes to seed storage biology, ERH is the preferred method over MC since it consistently affects decomposing organisms in all commodities, regardless of composition. This is even though there is a strong association between the moisture content of seeds in a particular setting and their ERH, which indicates the product's water activity (Chen, 2001). Farm-saved seed is stored in suitable containers in the coldest, driest area feasible after being allowed to naturally dry in the sun or wind (sun or shade drying). If the maximum ambient RH and temperature do not exceed 70% and 30 °C, respectively, for more than three months of the year, then the seeds of most grain crops can be safely stored under uncontrolled conditions for at least one crop season with the appropriate plant protection measures (Agrawal, 1982).

Management of Temperature

Both Harrington's thumb guidelines and Robert's viability estimates acknowledge the distinct and well-documented impacts of temperature and humidity on seed longevity. Subsequent research, however, has shown that humidity in the storage environment has a greater impact on seed survival than temperature. Seeds dried to a reasonable moisture content (MC) will last far longer than seeds kept at a higher moisture content, even at temperatures exceeding 25 °C. On the other hand, high moisture seeds are not suitable for subfreezing conditions; they should only be stored at or below 10 °C. Seeds with a moisture content of 12–14% deteriorate more physiologically and pathologically at temperatures above 25 °C, while seeds with a high moisture content (MC) may freeze-damage at subfreezing temperatures. In a study,

Wang *et al.* (2018) investigated the storability of primed rice seed (dry to 10.5% moisture) under various temperature, relative humidity, and air availability settings. The findings demonstrated that primed rice seeds could be cultivated in vacuum (V) at low temperature (LT <5 °C), room temperature (RT ~30 °C) and vacuum (RT-V), or in room temperature, aerated, and low humidity (LH 20–26%). However, under room temperature-aerated-high humidity (RT-A-HH) (60 °C and above), the viability of the seeds was significantly reduced. Therefore, when the seeds were stored in high relative humidity and aerated conditions, the negative impacts of storage temperature were clearly visible. The lifetime of primed seeds maintained in a vacuum was not affected by a temperature increase of 30 °C in a brief storage study, underscoring the critical role that air—or lack thereof—plays in determining how long seeds may be preserved.

Gaseous Environment of the Storage Area

It is well known that seeds, when dried to low moisture content (<5%) and stored in hermetically sealed containers (such as glass, metal cans, laminated and poly-lined/polythene, etc.), can be stored for long periods of time at ambient, lower, or higher temperatures (Grabe and Isley 1969). As opposed to seeds that are packaged in unsealed receptacles. The impact of gaseous components in the storage environment on the longevity of seeds stored in hermetically sealed containers has been the subject of numerous research (Bennici *et al.* 1984). There have been initiatives to extend the life of seeds, with varying degrees of reported success. CO2, vacuuming, and other inert gases are used in these procedures. Groot *et al.* (2015) claimed that airborne oxygen hastens the aging process of dry seeds and recommend storing seeds in anoxia to prolong germplasm preservation.

Storage Structures

The moisture level of the seeds, the kind of storage container utilized, and the storage atmosphere all have a significant impact on how long seeds stay in storage. In general, these elements combine to impact the physiological and biochemical characteristics of seeds that have been kept, which results in a decrease in seed quantity and quality, particularly in tropical and subtropical regions. Until they are required for the subsequent sowing, large quantities of seed are kept in storage. After harvesting, the seed crop is occasionally kept for a brief while in drying sheds or piled high in sizable stacks covered with waterproof materials like tarpaulin, until it is threshed, conditioned, and stored. But after the seed has been dried and processed, it must be kept in large quantities in buildings that are best built with waterproof roofs and walls, as well as with sealable holes for regular fumigation and regulated ventilation. It does not let water leak through the walls or the floor. Columns should be widely spaced, and ceilings should be high enough to allow seed bags to be stacked up to five meters high, to facilitate the use of forklifts and bag transportation in stores. For maximum reflection, seed warehouses with metal roofs usually need to be painted white and have enough insulation. When seeds are kept in large quantities, heat builds up, thus it's important to move or rotate the seeds occasionally to get rid of any hot patches. In stores, aeration using pedestal or moving fans helps dissipate heat. Several sensors are positioned at various locations inside modern seed storage facilities to monitor the temperature and humidity levels. According to Desai *et al.* (1997), these sensors can chill the seeds and start aeration as needed. Additionally, it is crucial to guarantee that the seed material in storage is nearly devoid of viruses and pests that can contaminate seeds. Rat traps are placed at possible access points to keep rodents out of the way, and seed warehouses in tropical and subtropical regions are built at least one meter above the ground. Seed warehouses with metal roofs often require adequate insulation and white paint for optimal reflection. Heat builds up when seeds are stored in big quantities, thus it's critical to periodically rotate or move the seeds to remove any hot spots. Aeration with pedestal or moving fans aids with heat dissipation in stores. Modern seed storage facilities have multiple sensors placed at different angles to keep an eye on the humidity and temperature. These sensors can cool the seeds and initiate aeration as necessary, according to Desai *et al.* (1997). Furthermore, it is imperative to ensure that the seed material in storage is almost completely free of pests and viruses that can contaminate seeds. It is recommended that the seed be arranged in bags and sacks on wooden pallets, keeping a safe distance from the walls and ceiling. To preserve seed quality, commercial stores consider various storage systems based on the type of seed and its intended application.

1. **Storage under ambient conditions with or without ventilation:** Commercial-scale unconditioned seed storage buildings rely mostly on natural ventilation and cooling to reduce the detrimental effects of heat build-up. In temperate climates, bulk seed storage with controlled natural ventilation is safe; but, in warm, humid climates, outside air ventilation might be harmful (Copeland and McDonald 2001). If kept in mild conditions, seed sacks, bags, piles, single layers, and open containers can be kept for a short period of time, like during pre-processing or before final packing and storage, or for 6–8 months after harvest until the next cropping season. Care is taken to ensure that rodents and birds stay away from the seeds.

2. **Storage in moisture controlled condition:** The ideal places for this kind of storage are those with moderate temperatures and persistently cool winters. There are two ways to manage the moisture content of seeds: either pack pre-dried low moisture (6–8% moisture) seeds in vapour-resistant bags or containers or put seeds in vapour-permeable sacks or bags and use dehumidifiers. Since controlling storage temperature in warmer regions can be very difficult, it is advised to use in-the-bag desiccants and frequently use dehumidifiers, especially during the wet season, to maintain the RH of the storage environment storage for at least one season and to carry over the unsold stocks for the next season.
3. **Storage in low-temperature conditions:** Low-temperature storage modules are often maintained by seed companies that deal with a broad variety of crop seeds in warmer climates. Low volume, high-value vegetable and flower seeds that are not suitable for storage are dried to a low moisture content (MC), sealed, and packed in containers that are resistant to vapour and moisture and stored in cold storage units (modules). However, seeds stored at low temperatures in pervious moisture containers require extra care. The RH of the surrounding air rises with decreasing temperature, raising the seed MC. When seeds are exposed to warmer temperatures, they will deteriorate more quickly after being stored in this manner.
4. **Storage under controlled conditions of RH and temperature:** Large amounts of seeds are kept in this type of storage for at least two or three years. Only high-value seeds (like hybrids); inbred parental lines of the hybrids, especially when the hybrids are twice, or triple crossed (like maize); or exceptionally poor storers (like onions) can benefit monetarily from it. This is also widely used for most vegetable seeds, especially small seeded or hybrid types, and flower seeds because of their high value per unit weight and smaller mass. According to Copeland and McDonald (2001), 20°C and 50% relative humidity are typically maintained. For operation, this type of shop requires a consistent supply of electricity.
5. **Moist cold storage with control of temperature for limited period storage of desiccation-sensitive seeds from temperate regions:** Refractory seeds with a high MC can be stored at low temperatures, just above freezing, for research purposes. The storage is maintained at almost saturation relative humidity, and the seeds are periodically moistened. This is rarely how seeds are stored for commercial use.

Types of Storage based on Storage Period

1. **Commercial seeds:** For one season, or eight or nine months from harvest to planting, the maximum number of seed is kept in storage. Commercial seeds are kept in all-purpose stores for a limited period after being dried to a moisture content of ≤14% for oilseeds and ≤14% for cereals.

2. **Carry over seeds:** In the main season, it is the unsold seed. In the primary season, between 20 and 25 percent of seeds are unsold. It needs to be stored carefully for a full year, or until the next time it is planted. Properly dried carry-over seeds are kept in airtight metal containers or moisture-proof bags at low temperatures.

3. **Stock seed:** It is a small amount of any variety of seed that has been developed by a plant breeder. Genetic drift could cause the variety to degenerate if it is regularly multiplied to preserve it. For this reason, a dehumidifier is used to keep the stock seeds in polythene bags with a gauge thickness of at least 700 and a relative humidity of 25% at a temperature of 20°C.

4. **Gene bank:** A seed gene bank is defined as an archive of cultivated plant germplasm resources and related species. Monitoring genetic erosion, or the progressive loss of genetic diversity, is crucial. With 0.4 million accessions, the National Bureau of Plant Genetic Resources (NBPGR) is the first global seed gene bank out of 1750. To maintain normal genetic variation within the population, a minimum of 4000 seeds for homogenous accessions and a minimum of 12000 seeds for heterogeneous material are needed for storage. The harvested seed is thoroughly dried, cleaned, and viability tested before being stored at a low temperature. The accession is initially kept in liquid nitrogen at a temperature of -196oC for an extended period. and a second medium-term storage location. (4–12°C up to 35% RH) is duplicated periodically, that is, once every three to four years, and utilized for distribution, evaluation, and regeneration. To avoid allowing water vapor to enter the seed container, the relative humidity inside should be airtight. Aside from seeds that have been reduced in moisture content, low ethylene and low oxygen concentrations are the most important variables for maintaining the viability of the seed.

5. **Storage at low temperature:** Low temperatures, such as -18°C to -20°C, work well for storing adequately dried seed for an extended period. The ideal temperature range for ultra dry orthodox seed storage is between

-5 and +5 degrees Celsius. A glass jar is used to keep the seed. Because CO2 is heavier and tends to occupy the lower regions of a volume, it can be used to replace the air in glass jars and decrease the presence of oxygen by packing as much seed within as feasible.

6. **Cryopreservation:** Cryopreservation is the process of storing seed at an extremely low temperature—-196°C, or the temperature of liquid nitrogen—for an extended period while stopping all biological processes. Since all enzyme functions are essentially stopped in these circumstances, biological material of any kind can be stored for an essentially endless amount of time. Because of the early infrastructure, operating costs, and creation of deadly intracellular ice crystals, it is not very well-liked. Procedures for cryopreservation can be divided into two categories. First, traditional gradual freezing; second, vitrification.

   1) **Slow freeze procedure:** Water in a cell's cytoplasm crystallizes into ice crystals at freezing temperatures. A cell becomes ragged and dies when ice forms inside of it. To prevent cell damage, the fluid inside the cell must be evacuated before the temperature drops below freezing. It is imperative to inhibit intracellular ice crystal formation during cryopreservation to ensure effective cell preservation. A variety of antifreeze (cryoprotectant) fluids can be employed to remove moisture from the cell prior to its freezing. After the seed has been effectively dehydrated, it is triggered by a concentrated mixture of osmotic agents and/or chemicals like glycerol, proline, abscisic acid, propanediol, dimethyl sulfoxide, ethanol, and sugars like sucrose and trehalose, among others. To gradually allow the cell to acclimate to the cryoprotectant while releasing water, the seed is exposed to it in a general stepwise manner. Once the moisture has been extracted from the seed cells, the seed and cryoprotectant are put into a glass container. The container is put in a freezer, which progressively drops the temperature. Following a brief period of freezing, seeding is carried out by touching the container with a tool—such as a pair of forceps—that has been previously cooled in liquid nitrogen. This initiates the creation of starter ice crystals in a single location that is apart from the cells. The remaining cooling ramps can start after the seeding process is finished. The container containing the cells can be submerged straight into liquid nitrogen to finish chilling it down to -196°C once it reaches temperatures between 30 and -85°C.

   2) **Vitrification:** Vitrification is the physical process of rapidly lowering the temperature to prevent intracellular ice crystallization in cells

and tissues during ultra freezing by converting the cytosol's aqueous solution into an amorphous glassy state. Because it prevents tissue damage, tissues can be stored for an extended period at -196°C without losing their viability. The process of vitrification proceeds so quickly that there is not enough time for ice to form, causing the cellular fluid to be transformed into a vitreous or glass phase without causing any harm to the cells. Another method for achieving vitrification is to partially dehydrate the material that will be sown on silica before directly dipping it in liquid nitrogen to retain it indefinitely without causing any genetic alterations. An alkaline glass vial filled with adequately dried seed is kept cold. When heated, it quickly softens without breaking. An accession number-containing label is inserted within the container. The vial contains dehydrated seed. A piece of dehydrated, hydrophobic silica gel is placed to each vial of seed. Using a rubber stopper, the tube is sealed and left for 15 to 20 days. Should this happen, the silica gel's colour in the vial containing the improperly dried seed will change. The contents in the glass vial need to be adequately dried. About 5 cm of the vial's top are left empty at this point. By heating, the vial is sealed. The sealed vials are immersed in water for 24 hours to check the seal. A set of 25 vials are placed in one jar for long term storage.

3) **Storage of seed under vacuum:** Dried seed storage is made easier with the development of a hand-operated vacuum packing method. The technology offers a way to seal containers that have less oxygen in them. A closed container keeps seeds from rehydrating, extending their shelf life. Insects kept with the seeds are suffocated by the low oxygen content, which also prolongs the seeds' viability by reducing the rate of seed respiration. The seed is kept at a low temperature when it is packaged.

Types of Storage

Storage of rice: Usually, farmers use rope made from paddy straw to build a structure that resembles a bin for storage. A thick layer of rice straw forms the base of the bin. This bin has rice seeds that have been properly dried. Rice straw is used to cover the filled rice bin, which is tightly formed into a cone.

Storage of maize: The seed of maize are stored by unpeeling of cob with the tip downward to prevent moisture entry and is stored by stacking/hanging on the rope or stick. This technique is performed throughout the world.

Underground: In the central region of India, underground storage of grain and seed is a widely used technique for rice, wheat, sorghum, and pearl millet. The location of the subterranean structure preparation shows no indications of water retention. A shallow well is dug, and a thick bed of straw is placed at its foot, with dry straw covering the well's perimeter. The subterranean structure is filled with properly dried seeds. Following the seed filling, a thick layer of straw is applied to the building once again, and then mud, cow dung, and straw are added last. Because of the high levels of carbon dioxide and low levels of oxygen, the seed is shielded from environmental changes, insect infestations, and mold growth in the underground storage.

Mud jar: A mixture of clay, straw, and cow dung cake is created in a huge jar with a hopper at the top and an exit at the bottom. The jar is filled with the grain or seed that has been well cleaned and dried, and the apertures are securely closed. The jar is set up either outside or in the storage area. Periodically, the grain or seed is taken out of the outlet based on demand. Before being sown, the topmost seed descends to the bottom and is utilized as seed.

Bagged: Seed is stored in gunny bags and stored as stack in the room by the farmers. The seed is vulnerable to environmental fluctuations and infestations of insect pests.

Farm bins: There are three reasons why farms need storage space: to store the crop right after harvest, to shift it to a better storage facility, or to sell. Smaller lots of grain preserve quality better than bigger lots when stored in bins. When grain is kept out of the hands of rodents, birds, insects, or dampness, farm storage often preserves grain quality better than elevator storage. Utilizing burnt clay, which has been used by farm families for ages, farmers created the oriental bins.

On farm Safe Seed Storage

Over 50% of the seed used in tropical and subtropical regions is kept on farms, where it is first air-cooled and sun-dried before being stored in hygienic buildings or containers in a cold, dry environment. For this, a variety of containers can be used, including cloth bags, jute/hessian bags, clay pots, metal bins, bamboo baskets with mud plaster, and HDPE (interwoven) bags. Mud-plastered structures are also used for larger amounts of grains and seeds. Many plant parts with insect-repelling properties can be used to create these structures/containers (Francis *et al.* 2015). Unless the seed is further packaged in an environment that keeps moisture vapor from entering, it is typically insufficient to just dry the seed for safe storage in a humid environment (such as moisture vapor impervious containers). It can occasionally be difficult to dry seed below 12% MC, especially for crops grown during the wet season that are harvested when the surrounding relative humidity is still very high (~70%). For storing high-volume cereal crop seeds at 13–14% mechanical moisture content, bags made of sheets impermeable to gaseous exchange or moisture-vapor-impermeable bags are useful. Storage methods such as Grain ProTM Super bags or hermetic cocoons (De Bruin *et al.* 2012) can be used for one to two seasons because they are moisture-proof and impermeable to both air and moisture (www.grainpro.com). These bags, together with larger-scale cocoons, can significantly improve commodity storage when used properly (Afzal *et al.* 2017). Another inexpensive and highly promising technique for on-farm seed conservation is the use of zeolite beads, also known as "drying beads," inside seed bags or drums to safely dry seeds at room temperature while maintaining seed viability and protecting against storage insect pests (Kunusoth *et al.* 2012; Sultana *et al.* 2021). The zeolite beads can regenerate hundreds of times by heating them, making them reusable. Bradford *et al.* (2018) proposed the dry chain concept as a secure means of storing grains, seeds, and food items. Products in the dry chain are packaged in water-impermeable materials and dried to low MC. This is like the cold chain, where products, such as fresh foods, are stored and distributed at low temperatures.

Good Storage Practices as a Preventive Measure

To maintain proper storage hygiene, stores and supplies should undergo crucial inspections on a regular basis. Furthermore, in terms of cost and necessity, the chemicals must be the very last choice. Pesticides should not be applied excessively to goods that are maintained in storage. It consists of two parts: measures for in- and pre-storage. The dry chain method, which combines gas-tight containers and drying beads, offers safe seed storage for short- to medium-term periods at a relatively low energy consumption and cost, as refrigeration is not required. Farmers can effectively store seeds on their properties or maintain a sufficient supply of seeds in community seed banks by using this method, especially in humid regions (Dadlani *et al.* 2016).

Pre-Storage Preventive Measures

Preparation of Seed Stocks

- The seed stock shouldn't contain any shattered or damaged seeds. Prior to storage, ensure that the seed is dried to a moisture content (MC) of less than 10%, or ≤13% for paddy. Pulses may include insect infection from the field when adult insects emerge from the seed in the pre-processing hall before processing. Pulse seed should therefore be fumigated to stop insect multiplication or sun-dried to remove any internal contamination as soon as it is planted in the godown.

- Confirm that new harvests do not carry field infection in other crop seeds. If a live bug is found, fumigate the area.
- Mix premium-grade Malathion 5%D at 200 g/t of seed with deltamethrin (K-Othrin 2.5 SC) at 40 mg/kg of seed. Treating seed for use as food or feed is never appropriate.

Preparation of Seed Store/Shed

1. Get rid of all spills, webs, and debris from all the structures. Disinfect concrete floors and walls by using a backpack sprayer to apply 50 ml of either fenitrothion 50 EC or malathion 50 EC in 5 L of water for every 100 m^3 of floor surface. Moisten all surfaces to get rid of insects hiding in nooks and crannies.
2. If used for non-commercial purposes, fumigate old seed bags before reusing them. Alternatively, treat the bag surface area of 20 m2 with either 02 mL/L of deltamethrin or 10 mL/L of malathion. Dry out old seed bags after immersing them in hot water (above 50 °C) for fifteen minutes.
3. Spinosad (Tracer 45 SC) and Indoxacarb (Avant 14.5 SC) treated rice seed offers effective (0.10% and 0.13%, respectively) control of storage insects infesting rice seed for up to 12 months of storage under ambient conditions; in contrast, seed damage was 7.4% control in untreated lots.
4. Before storing seeds, if needed, apply the following pesticides to the wall or bare surface: Malathion (50 EC) at 10 mL/L of water and 5 l/100 m2.
5. 40 mg/L of water and 5 L/100 m2 of deltamethrin (K-Othrin 2.5 SC).
6. The application of pesticides will follow current national regulations. The Heating Process Heat treatment has been used to control the growth of pathogens and insects.

Solar Heat (Solarization)

When utilized to create sun heaters, dark fabrics or black plastic sheets can trap ambient heat and eliminate pests. In tropical and subtropical regions where midsummer temperature might exceed 40°C, it functions particularly effectively. When a solarization cover is applied, pulse beetles and other pest storage insects can perish at temperatures above 60°C in about two hours. Pulse beetles in pigeon pea seeds, at any stage, died when they were solarized in polyethylene bags at ICRISAT in Hyderabad and reached temperatures as high as 65 °C, according to Chauhan and Ghaffar's (2002) research. Moreover, seeds that had been solarized retained their resistance to insect damage after 41 weeks of storage in a seed store setting. The germination of seeds is not negatively impacted by solarization. On farms, this technique is particularly effective for storing seeds. On farms, sunning and sieving, or S & S, was tested. This is sifting after solarization to get rid of dead bug debris. After four months of storage, the results showed that S & S was just as effective when paired with 1.6% methyl perimorphs or 0.3% permethrin seed treatment. It has been demonstrated that solarising seeds in clear polythene (700 gauge) packets for six days, for three hours each day, is an effective treatment for reducing insect damage in the majority of National Seed Programme (NSP) centres in India.

Cold Storage

The seeds have a longer shelf life because low temperature storage (<20 °C) in the godowns prevents the growth and establishment of storage insect pests and slows down physiological seed breakdown. Most storage insects have their life cycle prolonged but not completely eradicated. The length of time insects can live in cold weather is significantly influenced by the relative humidity (RH) in the enclosed space. Less than 30% relative humidity (RH) is ideal for medium-term seed storage (4-5 years). The type of seed and its moisture content, the seed's initial health (field infestation, if any), the insect species and its stage, density, and distribution in the seed lots, the air temperature and relative humidity, the duration of the exposure period, and other variables all affect how effective insect control is in cold storage.

Controlled Conditioned Seed Storage

The hot, humid weather is ideal for insect activity. There are other effects on the seed's quality and shelf life. Consequently, a seed go-down with low relative humidity (<50%) and low temperatures (<20 °C) increases seed quality and prolongs shelf life. Cold storage technique can be used to keep low-volume, high-value (LVHV) seeds, but it is expensive. Large-scale seed storage for wheat, paddy, or other crops would not be economical. A cold storage facility with a dehumidifier is usually used to store carry-over seed stock for a duration of 1-2 years at <20 °C and <50% RH, or for 3-5 years at 15°C and 30% RH for the storage of vegetable and fruit seeds.

Fumigants

When introduced into items or confined spaces, fumigants are gaseous or vapour-forming chemicals that permeate into them at concentrations that are lethal to the pest species they are meant to remove. Aerosols, often known as smokes, fogs, or mists, are airborne particulates. One of the most important and useful characteristics of the fumigants is their capacity to penetrate the fumigated materials, neutralize the target organisms, and then disappear. To keep storage insects under control, a range of chemical fumigants is recommended. Here is a list of various fumigants and an explanation of their properties; however, most of these are no longer in use in most countries due to the risks they pose to the environment and public health. It has also been noted that many insects have developed resistance to several of these.

Hydrogen Phosphide or Phosphine (PH_3)

1. For specific applications, tablets containing aluminium phosphide (AlP) are sold under the names "Celphos, Quickphos," etc. It weighs three and a half grams of pellets (used against rats) and releases one-third of its weight in phosphine gas.
2. Ammonium bicarbonate, carbonate of ammonium, paraffin, and urea are other components. The chemical process is AlP + 2NH4 OC (O) NH2 + 3H2O = PH3 + Al (OH)3 + 4NH3 + 2CO2.
3. In the presence of moisture, CO2 diffuses from the tablet and reduces PH3's flammability.
4. Ammonia is a warning gas and reduces the chance of fire.
5. Garlic or carbide are comparable in scent to that of aluminium phosphide. It weighs heavier than air and is not very soluble in water. Although it is a practical and secure method of evolving gas, it is extremely combustible on its own.
6. More quickly do adults and larvae perish. On the other hand, killing the eggs and pupae is usually the most challenging. Quite long exposure periods (10 days) can be too much for eggs and pupae to handle.
7. Phosphorus has no effect on legumes and cereal seeds that have undergone one or two fumigations at quite high concentrations. A multilocation trial funded by the National Seed Project in India found that the viability and vigour of different agricultural seeds were unaffected by up to four repeat administrations.

8. As a result, phosphine is a fumigant that is effective at keeping pest insects in check while being stored, and it can be used provided the applicable laws permit it.

Pre and Post-Harvest Strategies for Disease Management

Seed storability and disease incidence during storage are strongly influenced by the quality and state of the seed at harvest. Because of this, it's critical to plan seed production for safe sites and appropriate seasons when the risk of developing serious illnesses from seeds is known to be minimal or non-existent. Pre-harvest applications of suitable pesticides or bio-control agents and crop harvesting at the proper maturity stage can also help maintain the quality of the seed throughout storage. Exposure of seeds to biotic and abiotic stresses affects their vigour and longevity, especially during seed maturation (Siti *et al.* 2019). When pre-harvest rains and illnesses cause grain seeds, which are frequently sold uncoated or uncoloured, they lose their sowing and market value. In places where diseases or discolouration are expected, a pre-harvest preventative spray may be applied. Govindrajan and Kannaiyan (1982) found that rice grain discolouration was decreased by pre-harvest copper oxychloride spraying. It was also discovered that rice seeds with higher phosphate and nitrogen content discoloured more; however, rice seeds with wider field spacing between plants discoloured less (Mishra and Dharam Vir 1991). According to Deka *et al.* (1996), rice grains can effectively have their discolouration decreased by spraying them with regular salt after using Maneb when they are still in the boot leaf stage. On the other hand, it is claimed that the darkening of the seed coat in several crops, mostly legumes like soybean and peanut, is an indication of oxidative reactions rather than a disease (Siao *et al.* 1980). Rainy season cultivation and harvesting of seed crops is thought to have a stronger fungal connection. High-humidity samples usually have a higher proportion of storage fungus (Indira and Rao 1968). Due to a variety of characteristics, including their chemical makeup, the presence of alkaloids and antifungal chemicals in the seed coat, and others, seeds of different species and types have varying propensities to harbour storage fungus (Mishra and Kanaujia, 1973). Because of their thick, hard seed coat, *Luffa acutangula* seeds absorb less moisture during storage, according to Nair's (1982) assessment of the quantity of fungus on the seeds. Sheeba and Ahmed (1994) found that high-yielding, fertiliser-responsive paddy cultivars had varying susceptibilities to fungal assault during storage, and that fungal incidence on their seeds was higher than on traditional kinds. Relative humidity, storage temperature, and initial seed wetness all have a role in preserving seed health and germination. Thus, harvesting the seed at the appropriate stage of maturity is the first and most crucial step in maintaining its future storability. Because mechanical injuries promote the invasion of pests and diseases, it is crucial to avoid any physical damage to the seed during harvesting, threshing, and other post-harvest handling activities. Lal *et al.* (1976) reported that on groundnut kernels, acetic acid and propionic acid were effective against *A. flavus* and *Curvularia lunata*, while on wheat and maize grains, *A. niger*, *A. flavus*, *Penicillium oxalicum*, and *Alternaria alternata* were effectively controlled by propionic acid and potassium metabisulphite. Propionic acid and sodium metabisulphite, according to Vaidya and Dharam Vir (1974), prevented Aspergillus and Penicillium species from growing on groundnut kernels. It is essential to maintain the seed godowns dry, clean, sterilised, devoid of cracks and fissures, with adequate ventilation, and high-quality seeds when storing them protecting the seed against microbial exploitation both in storage and when it is planted in the field. The seed material must be packed into sanitary containers. Sometimes grain seeds can be bulk packed with fabric sacks and gunny (hessian). These must be adequately fumigated or disinfected to avoid infection by the carry-over pathogens. When any pests are discovered, they should be promptly removed using effective management methods such fumigation. It is important to regularly check seed inventories for the emergence of pests. Consequently, disease and pest management in stored seeds requires the optimal storage conditions as well as the use of treatments (seeds, godowns, and bags) that do not jeopardize the health of seed handlers and consumers. Horticultural crop planting materials, such as stems, roots, leaves, tubers, corms, rhizomes, suckers, grafts, etc., can carry a range of illnesses due to their high moisture content. These might cause diseases that make them less valuable as plants. These propagules may also be prevented from developing in the field by the pathogens present in the soil. Most of these problems can be mitigated by using an appropriate seed treatment technology, which also results in less chemical waste and more efficient disease management than later stage spraying.

Long-Term Seed Preservation

Protecting biodiversity and preserving plant DNA for future use are the two main goals of long-term seed storage. Because of this, seeds are maintained at extremely low MC and low to very low temperatures in the gene banks. For plant genetic variety to be preserved, made accessible, and used in agriculture, gene banks are crucial. Preserving seed samples' viability, genetic integrity, and quality is the primary objective of gene banks, as it keeps them usable even after decades in storage (FAO 2014). Thus, seeds are stored at low to subfreezing temperatures and dried until glassy to delay germination. When germination falls short of the targeted levels, stocks are refilled. Hay and Whitehouse (2017) suggested that instead of basing the regeneration of seed stocks on initial germination, these could be based on seed storage experiments to identify which seed lots to test first. They also suggest using tolerance tables in conjunction with sequential testing schemes to reduce the number of seeds used for viability testing. There are a few different methods for drying seeds, such as using a dehumidifier chamber or storing seeds with desiccants to create an equilibrium in a dehumidified atmosphere. The methods chosen will be influenced by several factors, including the equipment that is available, the size and amount of sample that need to be dried, the local climate, and budgetary considerations. However, there is a limit to how much lifespan may be extended through drying. It is therefore recommended that, for long-term preservation, seed samples be dried to an equilibrium of 10–25% RH in a regulated environment of 5–20°C (Cromarty *et al.* 1982), so the MC of seeds approaches for medium-term storage, samples are maintained between 0 and 10°C (10–15 years). Furthermore, working collections can be stored for three to five years at 5 to 10°C. These collections are used to assess, multiply, and disseminate the accessions for use. Usually, the base collection is periodically replenished, or the seeds of their accessions are multiplied to preserve active collections. Ultra-drying reduces seed moisture to 1-3 percent by using hot drying, freeze-drying, or dehydrated forced air drying. Kong and Zhang (1998) demonstrated that there was nearly no difference in longevity when seeds were dried on silica gel by freeze-drying or heating to 50ºC, provided that the seeds were not over-dried below 1.5%.

References

Afzal I, Bakhtavar MA, Ishfaq M, Sagheer M, Baributsa D (2017) Maintaining dryness during storage contributes to higher maize seed quality. J Stored Prod Res 72:49–53. https://doi.org/10.1016/j.jspr.2017.04.001

Agrawal PK (1982) Viability of stored seeds and magnitude of seed storage in India. Seed Tech News 12(1):47

Beninci A, Binoti MB, Floris C, Gennai D, Innocenti AM (1984) Ageing in Triticum durum wheat seeds. Early storage in carbon dioxide prolongs longevity. Environ Exp Bot 24:159–165

Bradford KJ, Dahal P, Van Asbrouck J, Kunusoth K, Bello P, Thompson J, Felicia W (2018) The dry chain: reducing post-harvest losses and improving food safety in humid climates. Trends Food Sci Technol 71:84–93

Chauhan Y, Ghaffar MA (2002) Solar heating of seeds - A low-cost method to control bruchid (Callosobruchus spp.) attack during storage of pigeonpea. J Stored Prod Res 38:87–9.

Chen C (2001) Factors which affect equilibrium relative humidity of agricultural products. Trans Am Soc Agric Eng 43(3):673–683

Copeland LO, McDonald MB (2001) Principles of seed science and technology, 4th edn. Kluwer Academic, Hingham, MA Cromarty AS, Ellis RH, Roberts EH (1982) The design of seed storage facilities for genetic conservation. International Board for Plant.

De Bruin T, Navarro S, Villers P, Wagh A (2012) Worldwide use of hermetic storage for the preservation of agricultural products. In: Navarro S, Banks HJ, Jayas DS, Bell CH, Noyes RT, Ferizli AG, Emekci M, Isikber AA, Alagusundaram K (eds) Proc 9th. Int. Conf. on Controlled.

Deka B, Ali MS, Chandra KC (1996) Management of gr[ai]n discoloration of rice. Indian J Mycol Plant Pathol 26:105–106.

Desai BB, Kotecha PM, Salunkhe DK (1997) Seeds Handbook. Marcel Dekker, Inc., New York, pp 531–545.

Dharam Vir (1974). Study of some problems associated with post-harvest fungal spoilage of seeds and grains, In: S.P. Raychaudhury and J.P. Verma (eds), Current Trends in Plant Pathology, pp. 296-304, Botany Department, Lucknow University, Lucknow.

Ellis RH (2019) Temporal patterns of seed quality development, decline, and timing of maximum quality during seed development and maturation. Seed Sci Res 29(2):135–142

Ellis RH, Hong TD (2007) Seed longevity—moisture content relationships in hermetic and open storage. Seed Sci Technol 35(2):423–431.

Ellis RH, Hong TD, Roberts EH (1989) A comparison of the low-moisture-content limit to the logarithmic relation between seed moisture and longevity in twelve species. Ann Bot 63(6): 601–611

FAO (2014). Genebank standards for plant genetic resources for food and agriculture. Rev. ed. Rome Fields PG, Muir WE (1996) Physical control. In: Subramanyam B, Hagstrum DW (eds) Integrated management of insects in stored products. Marcel and Decker, New York, pp 195–221

Francis O, Ogu E, Ikehi M (2015) Use of neem and garlic dried plant powders for controlling some stored grains pests. Egypt J Biol Pest Control 25:507–512

Grabe DF, Isley D (1969) Seed storage in moisture resistant packages. Seed World 104(2):4

Groot SPC, de Groot L, Kodde J, van Treuren R (2015) Prolonging the longevity of ex-situ conserved seeds by storage under anoxia. Plant Genet Resour 13:18–26

Harrington JF (1972) Seed storage and longevity. In: Kozlowski TT (ed) Seed biology, vol 3. Academic Press, NY, pp 145–240

Harrington JF (1973) Biochemical basis of seed longevity. Seed Sci Technol 1:453–461

Hay FR, Whitehouse KJ (2017) Rethinking the approach to viability monitoring in seed gene banks. Conserv Physiol 5: cox 009. https://doi.org/10.1093/conphys/cox1009

Indira K, Rao JG (1968) Storage fungi in rice in India. Kavaka 14:67–76

Justice OL, Bass LN (1978). Principles and practices of seed storage. USDA Agricultural Handbook, 506

Khare, D and Bhale, M.S (2019). Seed Technology (2nd revised & Enlarged Edition). Scientific Publishers, 5A, New Pali Road, Jodhpur, 342001 (India).

Kong X-H, Zhang H-Y (1998) The effect of ultra-dry methods and storage on vegetable seeds. Seed Sci Res 8(1):41–45

Kunusoth K, Dahal P, Van Asbrouck JV, Bradford KJ (2012) New technology for post-harvest drying and storage of seeds. Seed Times (New Delhi) 5:33–38

Lal SP (1976) Studies on storage fungi of wheat and maize. Ph.D thesis, IARI- New Delhi

McDonald MB, Nelson CJ (eds) (1986) Physiology of seed deterioration. Crop Science Society of America, Madison, WI

Mishra AK, Vir D (1991) Assessment of losses due to discoloration of paddy grains. I. Loss during milling. Indian J Mycol Pl Pathol 21:277–278

Misra RR, Kanaujia RS (1973) Studies on certain aspects of seed borne fungi II. Seed borne fungi of certain oilseeds. Indian Phytopath 26:284–294

Nair LN (1982) Studies on mycoflora of seeds of some cucurbitaceous vegetables. J Indian Bot Soc 61:343–345

Omobowale MO, Armstrong PR, Mijinyawa Y, Igbeka JC, Maghirang EB (2016) Maize storage in termite mound clay, concrete and steel silos in the humid tropics: Comparison and effect on bacterial and fungal counts. Trans Am Soc Agric Biol Eng 59(3):1039–1048. https://doi.org/10. 13031/trans.59.11437

Pritchard, Hugh & Dickie, John. (2003). Predicting Seed Longevity: the use and abuse of seed Viabilityequations; https://www.researchgate.net/publication/249968798_Predicting_ Seed Longevity_the_use_and_abuse_of_seed_viability_equations Ranganna S (1986) Manual of analysis of fruits and vegetable products. Tata McGraw Hill Publishing Co. Ltd., New Delhi, pp 441–495 Govindrajan K, Kannaiyan S (1982) Fungicidal control of grain infection. Int Rice Res News 7:1

Pérez-García F, González-Benito ME, Gómez-Campo C (2007) High viability recorded in ultra-dry seeds of 37 species of Brassicaceae after almost 40 years of storage. Seed Sci Technol 35:143–153

5

Assessing Physical Physiological and Chemical Changes During Seed Storage

Sharmila Dutta Deka and Priyanka Sharma

Concept: It's a natural process that modifies the cytology, physiology, biochemistry, and physical characteristics of seeds. These modifications lower viability and ultimately result in the seed's death. A decline in the proportion of germination indicates degradation, and weak seedlings are produced by seeds that do germinate. The factors that impact the ability of seed to be stored include the seed's quality at the time of storage, its pre-storage history (environmental factors during pre- and post-harvest stages), its moisture content or ambient relative humidity, the temperature of the storage environment, the amount of time the seed is stored, and biotic agents. Seed will inevitably get damaged when being stored (Balesevic-Tubic *et al.*, 2005). The quality of the seed at the time of storage, its pre-storage history (environmental factors during pre and post-harvest stages), its moisture content or ambient relative humidity, the temperature of the storage environment, the length of time the seed is stored, and biotic agents are factors that affect the seed's ability to be stored. Different physical changes can happen to seed during storage because of factors like temperature, moisture content, and length of storage. The viability and vigor of the seeds may be impacted by these modifications.

Physical Changes

1. **Moisture Content:** Seeds tend to lose moisture during storage, a process known as desiccation. Excessive moisture loss can lead to seed dormancy or even death, while too much moisture can promote fungal growth and seed deterioration. Maintaining optimal moisture content is crucial for seed longevity. One key component affecting a seed's viability during storage is its moisture level. The rate of seed deterioration increases with increase in moisture. High moisture stimulates physiological activity in storage, just as it is necessary for germination in the field.

Storage is basically a period of suspended growth, and the seeds must be kept at minimal levels of physiological activity at which the life can be maintained. Before storing, seeds must be dried to safe moisture levels because their moisture content determines how long they will last. The length of storage, the style of storage structure, and the type of seeds to be stored all influence the safe moisture content, though.

2. **Temperature Effects:** Temperature plays a significant role in seed storage. High temperatures can accelerate seed aging and reduce viability, while low temperatures can slow down metabolic processes but may also increase the risk of chilling injury. Temperature fluctuations can also lead to condensation within storage containers, which can promote fungal growth and seed decay. One of the most significant environmental elements influencing the viability and vigor of seeds during storage is the temperature. The duration of the germination capability will increase with decreasing temperature. As a result, refrigeration, insulation, or ventilation should be used to regulate temperature.

3. **Oxygen Levels:** Oxygen is necessary for seed respiration, but high levels of oxygen during storage can promote oxidative damage and reduce seed viability. Vacuum sealing or using modified atmosphere packaging techniques can help mitigate the effects of high oxygen levels. Increased oxygen gas pressure while storage shortens the seeds' lifespan by boosting their biological activity. As a result, to improve storability, biological activity in the seed is reduced during storage, and low oxygen pressure is attained by raising the atmospheric concentrations of nitrogen and carbon dioxide. Storage of seed in vacuum or in the environment of nitrogen is based on reduction of biological activity during storage.

4. **Physical Damage:** Mechanical damage during handling and storage can occur, leading to reduced viability. Seeds can be crushed, cracked, or bruised, especially during packaging and transportation processes. Proper handling and packaging techniques can minimize physical damage.

5. **Seed Coat Integrity:** The integrity of the seed coat is essential for protecting the embryo from external stresses. Damage to the seed coat can occur due to physical abrasion or microbial activity, which can compromise seed viability.

6. **Pest and Pathogen Infestation:** Seeds are susceptible to infestation by pests and pathogens during storage. Insects, rodents, and fungi can all damage seeds and reduce their viability. Proper sanitation, pest

control measures, and storage conditions can help prevent infestation. Pathogen, insect pests and other organisms which feed on stored seed cause significant damage to the seed by eating, injuring, damaging, and increasing heat or moisture that stimulates attack of mold, mites, and other pathogens. The organisms associated with deterioration of seed in storage are:

Fungi: Two types of fungi viz., field fungi and storage fungi attack the seed in the storage. Primary infection of field fungi is always during development or before harvesting in the field, but never in store viz., Alternaria, Fusarium and Helminthosporium spp. Whereas, the primary infection of storage fungi viz., Aspergillus and Penicillium is always during storage of seed. Infection of fungi decreases the viability by producing mycotoxins that develop stuffiness and clumps due to enhanced temperature. Storage fungi never grow at low level of moisture percentage i.e., 13% in cereals and 7-8 % in pulses and oilseeds at relative humidity below 68% temperature and enhanced metabolic activity.

Insects and mites: Insect and mite-induced seed deterioration is a major issue, especially in warm, humid regions. Insect pests viz., weevils, flour beetles, borers etc. deteriorate the seed with increase in moisture content (≥15%) and temperature (≥30%). The activity of these insect pests is checked at low moisture content (≤8%) and temperature (≤20%). Activity of mites is associated with high relative humidity (≥60%), but not with extremes of temperature. Pre harvest sanitation and spray in pulses before harvesting to control the insect infestation of bruchid are recommended.

Rodents and birds: Rodents are serious problems of storage that may result into a complete loss of seed. Entry of rodent in the store can be prevented by elevating the floor by 90 cm height above the ground level with escorting of 15 cm around the building at 90 cm height. Birds are also responsible for huge loss of seed in poor storage condition. All openings including ventilators should be screened/ sealed to prevent the entry of birds in the store.

7. Germination Rate: Over time, seeds may experience a decline in germination rate, (Fig 1) even if viability is maintained. This can be due to factors such as ageing, genetic factors, or physiological changes within the seed.

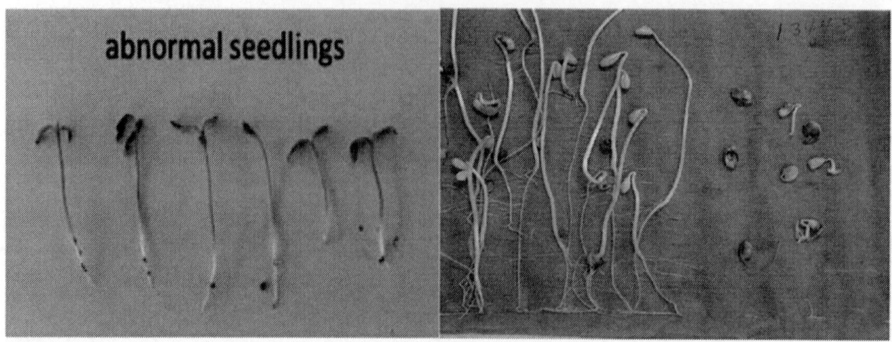

Fig. 1: Abnormal seedlings due to the physiological changes leading to decrease in germination rate

In Dicot: The embryonic axis' developing points, the shoot and root, are where degradation starts.

In Monocot: At the root tip, deterioration starts. reduces the amount of radicle extension more than coleoptile extension.

To mitigate these physical changes during seed storage, it is essential to implement proper storage techniques such as controlling temperature and humidity, using appropriate packaging materials, conducting regular monitoring for pests and diseases, and rotating seed stocks to ensure freshness. Additionally, using seed treatments such as priming or coating with fungicides can help maintain seed quality during storage.

Physical changes

During seed storage, various physiological changes occur that can affect the viability and germination capacity of the seeds. A few examples of the variables that affect these changes are temperature, humidity, oxygen content, and seed makeup.

Here are some of the key physiological changes that take place during seed storage

1. **Respiration:** Seeds respire even when dormant, consuming oxygen and producing carbon dioxide and water. During storage, respiration rates may decrease, particularly if the seeds are stored at low temperatures. However, excessive respiration can lead to depletion of stored reserves and reduce seed viability.
2. **Moisture content:** Moisture content plays a critical role in seed storage. High moisture levels can lead to fungal and bacterial growth, as well

as seed deterioration through processes such as hydrolysis and lipid oxidation. Low moisture levels, on the other hand, can induce seed dormancy or desiccation, which may reduce viability.

3. **Temperature effects:** Temperature influences seed metabolism and longevity during storage. Generally, lower temperatures slow down metabolic processes and reduce deterioration rates. However, extreme temperatures can still damage seeds. For example, freezing temperatures can cause ice crystal formation and physical damage to cellular structures.
4. **Seed dormancy and germination:** Some seeds have innate dormancy mechanisms that prevent germination under unfavourable conditions. During storage, dormancy may break down gradually, allowing seeds to become more responsive to germination cues such as moisture and temperature. Conversely, prolonged storage can induce secondary dormancy in some seeds.
5. **Loss of viability:** Over time, stored seeds gradually lose their ability to germinate and establish viable seedlings. Several variables, including the seeds' physiological age, storage circumstances, and genetic predisposition, can affect this loss of viability. It can result from various processes including ageing, oxidative damage, and deterioration of cellular structures.
6. **Metabolic changes:** Metabolic processes within seeds continue during storage, particularly at reduced rates. These processes involve the utilization of stored reserves such as carbohydrates, lipids, and proteins. Changes in metabolic activity can affect seed quality and viability over time, especially if reserves are depleted or metabolic by-products accumulate.
7. **Free radical accumulation:** Owing to oxidative stress, seeds may accumulate free radicals and reactive oxygen species (ROS) during storage. Reactive chemicals have the power to harm lipids, proteins, and DNA in cells, which lowers the viability and germination capacity of seeds. To mitigate these physiological changes and maintain seed quality during storage, proper management practices such as controlling temperature and humidity levels, using appropriate packaging materials, and periodic monitoring of seed viability are essential.

Chemical Changes

The capacity of conventional seeds to withstand desiccation and maintain viability in a long-dry state is one of their defining characteristics. But over time, as they age in storage, these seeds lose their ability to germinate. Numerous thorough analyses have determined that the main causes of seed aging include lipid peroxidation caused by free radicals, protein degradation or enzyme inactivation, disruption to the integrity of genetic material (nucleic acids) and disturbance of cellular membranes (McDonald, 1999). Most studies on biochemical breakdown during seed aging have used high temperatures and high seed water contents to simulate accelerated aging (McDonald, 1999). In a few days or weeks, seeds generally lose their viability under such storage conditions. Because of the low seed water content and chilly storage environment, seeds under long-term storage circumstances are likely to be in a glassy state. Many harmful activities may be prevented or inhibited by the seed cytoplasm's exceptionally high viscosity and poor molecular mobility (Williams and Leopold, 1989). If the temperature rises or the amount of seed water increases, the solid-like glassy state may soften into the rubbery state or even "melt" into the liquid state because the glass transition temperature (Tg) will drop below the storage temperature. Some deteriorative reactions that would ordinarily be slowed down in the glassy state could continue more quickly in the rubbery or liquid state due to their low viscosity and increased molecular mobility. Thus, depending on the seed cytoplasm's transition temperature, the main fundamental process that starts seed ageing may vary depending on the storage environment. During seed storage, several chemical changes occur, impacting the viability and germination potential of the seeds. These changes are crucial for maintaining seed quality over time.

1. **Respiration:** Respiration continues in seeds even after they are detached from the parent plant. Initially, respiration rates may be high, consuming stored reserves like carbohydrates and lipids. As storage progresses, respiration rates decrease, helping to maintain seed viability.

2. **Moisture content:** Seeds lose moisture during storage due to desiccation. Proper drying is essential to prevent microbial growth and seed deterioration. High moisture content can lead to mold growth and seed viability loss.

3. **Lipid oxidation:** Lipids in seeds are susceptible to oxidation, which can lead to rancidity and loss of viability. Factors like high humidity, high temperature, and oxygen exposure speed up this process. Antioxidants naturally present in seeds help to mitigate lipid oxidation. One crucial aspect of seed deterioration is the rupture of the plasma membrane, which is mostly brought on by lipid peroxidation. As seeds mature, higher ROS concentrations will target the membrane phospholipids' polyunsaturated fatty acids, causing the long-chain fatty acids to disintegrate into smaller

molecules and alter membrane permeability, ultimately resulting in membrane disintegration.

4. **Protein degradation**: Proteins are degraded over time, particularly when seeds are exposed to unfavourable storage conditions. Protein degradation can lead to loss of enzymatic activity and reduced seed viability.

5. **Carbohydrate metabolism:** Carbohydrates serve as energy reserves for seed germination. During storage, carbohydrates may undergo hydrolysis, converting complex carbohydrates into simpler forms that are more readily available for germination.

6. **Free radical formation:** Free radicals can be generated during storage due to various factors such as temperature fluctuations, exposure to light, and oxidative stress. Free radicals can damage cellular components and compromise seed viability.

7. **Changes in hormone levels:** Gibberellins and abscisic acid (ABA) are two important hormones that control when seeds germinate and remain dormant. During storage, changes in hormone levels may occur, affecting the dormancy status and germination potential of seeds.

8. **Metabolic pathways:** Overall metabolic activity in seeds decreases during storage. Metabolic pathways may shift, prioritizing processes essential for maintaining viability overgrowth and development.

9. **pH changes:** pH levels within seeds can change during storage due to metabolic processes and interactions with the storage environment. Maintaining optimal pH levels is important for preserving seed viability.

10. **Secondary metabolite accumulation:** Secondary metabolites such as phenolics and flavonoids may accumulate in seeds during storage. These compounds can act as antioxidants, protecting seeds from oxidative damage.

11. **Loss of Enzyme Activity:** Tetrazolium (TZ) and glutamic acid decarboxylase activity tests are the most sensitive methods for determining early seed degeneration. There was also a correlation found between seed deterioration and other oxidase enzymes, including catalase, peroxidase, amylase, and cytochrome oxidase.

12. **Increase in Seed Leachates:** One common indicator of deteriorating seeds is an increase in leachate content when the seeds are immersed in water. The soluble sugar content of the leachate can be ascertained by

measuring the concentration of the leachate using electrical conductance methods.

13. **Increase in Free Fatty Acid Content:** As seed moisture content increases, phospholipids hydrolyze more quickly, releasing glycerol and fatty acids in the process. The regular build-up of free fatty acids is harmful to regular cellular metabolism and results in a drop in cellular pH.

14. **Membrane Degradation:** It is widely acknowledged that one of the main reasons for loss of viability is a lack of cellular membrane integrity. Reduced membrane permeability during severe storage conditions causes more components of the seed to leach and ultimately results in a loss of viability. Electrolyte leakage is exacerbated by membrane disintegration during seed degeneration. Elevated levels of electrolyte leakage are linked to declines in seed germination, field emergence, and seedling vigor. Normal cell activity and energy production are diminished when membrane systems, such as the tonoplast, plasma lemma, and endoplasmic reticulum, are altered. Early in the degeneration of seeds, membrane degradation and permeability loss take place.

15. **Enzymes Alterations:** Changes in the activity of some enzymes, including DNase, dehydrogenase, lipase, ribonuclease, acid phosphatase, protease, diastase, catalase, and peroxidase. Scavenger enzymes like catalase and peroxides can neutralize ROS and hydrogen peroxides, which are created by a variety of metabolic processes. As as it ages, peroxide activity significantly declines. Because of this, seeds create lipid peroxidation products including lipid conjugants and monoaldehyde and become more susceptible to the impacts of oxygen and free radicals in membrane unsaturated fatty acids.

Changes in Cell Chemical Constituents

Studies have shown a large increase in free fatty acids and reducing sugars and a decrease in protein, oil content, and total sugars in degraded seeds. According to certain research, oligosaccharide, which has been linked to stabilizing membranes, reduced over time in storage.

1. **Reduced Metabolic Activity:** Elevated relative humidity accelerates the breakdown process and causes a decrease in nucleic acids during an extended storage duration. Compared to viable seeds, non-viable seeds have lower metabolic activity. The ability to synthesize nucleic acids and nucleotides is reduced by long-term storage.

2. **Free Radical Damage:** A portion of degradation is linked to the build-up of free radicals generated during metabolism. Seed storage exposes lipids to a steady assault by oxygen, which results in the formation of free radicals, hydrogen peroxides, and other oxygenated fatty acids. Since free radicals are erratic, they could react and harm molecules in the vicinity. In the absence of enzyme activity, oxygenated fatty acids would build up in the dry seed, harming cellular components and causing the seed to deteriorate. The two main reasons in which oil seeds deteriorate during storage are lipid peroxidation and the production of free radicals.

3. **Chromosome Aberrations:** One of the alterations associated with seed aging is chromosomal aberration, which is also referred to as mutagenic consequences. A few of the chromosomal changes found in seeds include nuclear size differences, ring formation, fusion, fragmentation, and bridges. Other reasons why things deteriorate are:

    - Functional structural degradation,
    - Biochemical alterations leading to decreased ATP levels,
    - A drop in sugar concentration,
    - The ribosomes' seed to separate,
    - Hydrolytic enzyme formation and activation,
    - Enzyme degradation and inactivation (amylase, dehydrogenase, oxidases, phospholipase, glutamic acid decarboxylase),
    - Meristematic cell starvation,
    - Decreased respiration, an increase in free fatty acid content and seed leachates, and the buildup of harmful substances.

Furthermore, the primary cause of seed deterioration, lipid peroxidation, results in the first biochemical alterations in seeds that are noticeable when they are in storage. The primary causes of the oil plant seed's quick deterioration are auto oxidation of lipids and a rise in free fatty acid content during storage (Balesevic-Tubic *et al.*, 2005).

References

Balesevic-Tubic S, Malenèiæ D, Tatiæ M and Miladinovic J (2005), "Influence of aging process on biochemical changes in sunflower seed", HELIA, Vol. 28, No. 42, pp. 107-114.

Biabani A, Boggs LC, Katozi M and Sabouri H (2011), "Effects of seed deterioration and inoculation with Mesorhizobium cicerion yield and plant performance of chickpea", Australian Jour. of Crop Science, Vol. 5, No. 1, pp. 66-70.

McDonald MB. 1999. Seed deterioration: physiology, repair and assessment. Seed Science and Technology 27, 177±237.

Williams RJ, Leopold AC. 1989. The glassy state in corn embryos. Plant Physiology 89, 977-981.

6

Principles and Methods of Seed Drying

Sharmila Dutta Deka and Priyanka Sharma

Concept: Reduction of seed moisture content to the acceptable levels for enhancement of storability and longevity is known as seed drying.

The moisture content of seed is negatively related to its storability. As the moisture content of seed increases, its storability decreases. Adequate drying of seed prolongs viability for reasonably long periods without cold storage. The process of drying begins with physiological maturity till storage/packaging of seeds to avoid deterioration. The seed is hygroscopic in nature therefore it adjusts its moisture content in equilibrium to the existing environment.

Seed Drying

Seed drying is the process in which an environment with low humidity is created to remove moisture from the seed up to the desired level without any damage to its viability and physical appearance. This seed is stored at low moisture content under the low relative humidity to enhance its longevity. The initiation of seed moisture content and the desired level are what dictate the amount of the drying process, and these elements are determined by the following

- **Type of seed:** This will consider the longevity behaviour, seed content, and seed structure (i.e., how well the species stores energy).
- The location's quantity of wind and sunshine, as well as the infrastructure and equipment needed to employ artificial drying profitably.
- **Conditions of storage to be followed:** The seed must be known to be stored either
  - (a) In vapour-permeable containers and left out in the open,
  - (b) In vapour-permeable containers and kept in stores with controlled humidity and temperature,
  - (c) In hermetically sealed containers and kept out in the open, or
  - (d) In hermetically sealed containers and kept in conditioned stores.

- **Value and volume of operations:** Other significant factors in commercial seed operations include the number of seed to be dried and the seed's worth per unit weight.

Drying Mechanisms

Individual seed: In a seed, water is present in bound form with amino or carboxyl group, absorbed from loose binding with hydroxyl and amide group and free form present in the tissues as capillary force. The moisture level of seed reaches 10-12% by removal of moisture present as free form in the seed. It can be achieved by reducing relative humidity under ambient temperatures. The basic mechanisms involved in drying of seeds are.

1. Due to their hygroscopic nature, seeds lose moisture until they reach equilibrium with the surrounding air. The air's temperature and relative humidity determine the equilibrium moisture content (EMC).
2. The free bound water in seed or in nearby atmosphere is always in exchangeable condition. The movement of water from seed to air is termed as evaporation, whereas the movement of water molecule from air to seed is termed as absorption. Both are continuous process in seeds. When evaporation exceeds absorption, seed dries out and when absorption exceeds evaporation, the seed takes moisture from atmosphere.
3. Because of the environment's high temperature and low relative humidity, moisture evaporation from the seed's surface enters the atmosphere.
4. The flow of air carries away the evaporated moisture that makes the low RH in the atmosphere.
5. The seed's surface receives moisture that was previously inside.

Seed lot: When seeds are dried by forcing air through them, not every seed dries at the same rate. During the process of drying, the seeds of seed lot can be divided into three zones.

Dried zone: Seed present near the inlet gets dried first and dry below the desired level earlier than the seed present away from the inlet. The moisture of seed and atmosphere in this region is at equilibrium.

Drying zone: The air passing through the dried zone reaches in the next region to carry moisture from the seed, until the moisture reaches equilibrium. This region is called a drying zone. The location of the drying zone is in between dry and wet zone. With the process of drying, the depth of dried zone is enhanced whereas wet zone is reduced. The active process of drying seeds occurs in drying zone.

Wet zone: Refers to the area above the drying zone, or the portion of the seed that is moist (16–20%) and situated between the top of the drying zone and the top surface of the seed.

The topmost layer of the seed in the wet zone contains maximum moisture whereas layer near the drying zone is driest to reach at equilibrium. Stratification is the term used to describe the variation in moisture content of the air entering and exiting the seed. The amount of air passing through the seed and its relative humidity determine how much stratification and how wide the drying zones are.

Factors Determining the Rate of Seed Drying

Seed: The seed's moisture content plays a crucial role in the drying process. The weight per volume, or bulk density, of seed is significantly impacted by its moisture content. The air flow, the depth of the seed, its size, shape, and moisture content, as well as the kind and amount of pollution (stones, straw, weeds, etc.), all affect how much energy is needed to move air through a bed of seed. Heat is used as energy to drive out the moisture in the seed. A seed's latent heat of vaporization is significantly higher than the latent heat needed for water to evaporate, depending on the temperature and moisture content of the seed. The rate at which moisture migrates from the seed's centre to its surface is determined by the seed's temperature, physical structure, chemical makeup, and permeability of its seed coat.

Air: One of the main factors influencing how quickly moisture is removed from the drying seed is the following characteristic of the airflow surrounding it.

Temperature of air: Hot air has capacity to carry more amount of water. As per Harringtons report, one kg air may carry 14,8 g of water at 20 °C, whereas at 40 °C, it enhances up to 41.4g.

Temperature °C	Capacity to carry water (g water/kg air)
0	3.9
10	7.6
20	14.8
30	26.4
40	41.4

Relative humidity in air: The amount of moisture removed from the surface is influenced by the drying air's relative humidity and surface saturation.

Velocity of air: Increased air velocity speeds up the drying process by assisting in the elimination of moisture from the seed's surrounding environment. At first, the moisture content drops quickly, but as the seed loses moisture, the pace of drying decreases.

Damage to Seed During Drying

Temperature of air: In general, more than 40ºC temperature of drying air may reduce the viability of seed. Drying at low temperatures maintains the viability of seed.

Low RH of air: Rapid drying with low relative humidity in air also damages the seed by developing cracks on the seed coat. Soyabean seed dried with the air of less than 40% relative humidity loses its viability even at low temperature.

Velocity of air: Drying under high velocity of air may develop stress cracks on seed because of unequal drying of individual seed due to quick evaporation of water from seed. Whereas, in slow drying seed deteriorates due to high humidity and temperature for longer period.

Damage Due to Excessive Drying of Seeds

After drying, the moisture content of the seeds should be suitable for storage. The desired moisture content depends on the type of seed, duration of storage, type of packaging and condition of store. Damages caused by excessive drying of seeds are

1. **Cracking of seed coat:** The internal kernel temperatures can be raised to the point that the endosperm breaks by drying in hot air or by spending too much time in the sun. The rate of drying has an impact on the degree of stress cracking. The development of stress cracks in seeds can also be attributed to rapid cooling.
2. **Cracking of seed:** Excessive drying may separate cotyledons from embryo.
3. **Nutritive value:** High temperature during drying may denature proteins, sugars, enzymes, and other complex structures of nutrients stored in the seeds. It affects vigor of seed drastically followed by formation of abnormal seedlings.
4. **Seed viability:** Drying under high temperature reduces the viability of seeds. Seed embryos are killed by temperatures greater than 40-42ºC.
5. **Appearance:** The colour and texture of the seed surface can be adversely affected by high temperature during drying of seeds.

6. **Handling of seeds:** Over dried seeds become fragile and susceptible to mechanical damage.

Methods of Seed Drying

Sun drying

The sun provides a significant and endless source of heat that helps seeds lose moisture. The wind's ability to extract the evaporated moisture speeds up the drying process. Sun drying is the most common drying method as it is cost effective, environmentally friendly with no damage to seed viability.

Stages of sun drying of seed: Usually, the moisture content of seed is decreased in the field prior to harvest and subsequently through sun drying on the threshing floor. Crops are harvested in the field when they are completely dry, and then the harvested produce is left in the field for a few days to dry in the sun. The winnowed and threshed produce is then spread out in a thin layer on threshing floors to finish drying in the sun.

From physiological maturity to harvestable maturity on plants: After physiological maturity, the seed has no association with the mother plant. The crop is left on the plant to dry the vegetative part of mature plant and loss of moisture from seed by evaporation.

From physiological maturity up to threshing on harvested plant: In central and north Indian states, kharif crop is harvested in the month of September and harvested plants are stored as stack or in the bundle up to sowing of Rabi crops. In double cropped area, just after sowing of Rabi crops, threshing of kharif crops starts. Currently, intensity of light is very high that helps in fast sun drying of seed on harvested plant. Stacking of harvested plants helps in movement of air through seed.

After threshing for storage: Drying threshed seed on flat exposed surface is the most common way of seed drying. Seed should be spread on black plastic sheet to get clean and uniformly dried seed. For large quantity of seed, a thin layer should be spread on elevated concrete floor. To facilitate uniform drying, the seed is raked 7-8 times a day. Drying seeds under shade in ambient condition is recommended for many crops viz., rice, soyabean etc. as it reduces the seed/ seed coat cracking and maintains the sowing seed quality. Covering the seed around midday is beneficial under hot and sunny conditions of summer. After drying, the seed is cooled down under low relative humidity before storage.

Forced air drying principle: The movement of moisture content of seed depends on temperature and relative humidity of surrounding air because of its hygroscopic nature. When the relative humidity of air is lesser than the seed, then moisture from seed will move out and the process is enhanced under high temperature and velocity of air.

Procedure: The recommended depth of the seed is inserted into the drying bin. Wet seeds are compressed with recommended-temperature air to extract the water. The air's temperature provides the heat required for the water to evaporate. The pace at which surface moisture evaporates in the surrounding air and the rate at which moisture migrates from the centre of the seeds to the surface determine how quickly seeds dry. After completion of seed drying, natural air without heat is forced in the seed lot for 30-60 minutes to bring the temperature of seed up to ambient level.

Types of Forced Air Drying

Natural air drying: Natural air of ambient temperature is used for drying seeds.

Drying with moderately heated air: The temperature of the air forced for drying is raised to 10 to 20°F to reduce relative humidity of air and enhance evaporation of moisture from seed.

Drying with high heated air: The temperature of the air forced for drying is raised to 110°F to reduce relative humidity of air and enhance evaporation rate from seed.

Heated air-drying systems: This involves construction of bins/storage structures for drying and air distribution system. The building for seed drying system depends upon the size of operation, number of crops to be dried and level of mechanization desirable. Air distribution systems for seed drying are:

1. **Main and lateral duct air distribution system:** In this system, the main duct is in the center or one side of the bin.
2. **Single central perforated duct system:** In this system, the main duct made of perforated metal is located in the center of the bin. The seed is filled around the duct up to equal thickness but not exceeding 6 feet. The air is forced upwards in the perforated main duct that flows laterally to dry the seed.
3. **Perforated false floor air distribution system:** The air is introduced in the bin through the perforated false metal sheet floor to dry the seed.
4. **Multiple storage bins:** To dry seed of many crops simultaneously, the seeds are stored in different bins and the flow of air from one source is controlled by sliding air gate of the respective bins.

Types of Forced Air Seed Dryers

1. **Layer in bin dryer:** The number of seeds, the drying unit's capacity, and the size of the bin determine how deep the bin should be filled. To dry the seed, heated air is pushed into the container. More seed is put to the next layer for drying when the seed has dried to the necessary moisture content. This technique ensures that the seed is dried evenly.
2. **Batch in bin drier:** Depending on the seed moisture content, the drying unit's capacity, and the size of the bin, the bin is filled to a particular depth. To dry the seed, heated air is pushed into the container. The seed is cooled and then put in a storage bin after drying. The following batch of seeds is transferred and then dried.
3. **Batch dryer:** The seed is placed in bin with inner air chamber surrounded by two parallel perforated steel walls that contain seed up to desired thickness. The heated air for drying and natural air for cooling are forced through the seed. After drying, the seed is removed and replaced by another batch.
4. **Continuous dryer:** In this method, flow of seed in bin for drying is continuous through feeder whereas dried seed is collected at the outlet. Heated air is forced through the upper 2/3 seed column.
5. **Wagon drying:** It is a type of batch drying in which seed is loaded from combine into the wagon with perforated floor for drying. The wagon relates to the air distribution duct to force the hot air through the perforations of the wagon floor for drying of seed. After completion of seed drying, natural air without heat is forced in the seed lot for 30-60 minutes to bring the temperature of seed up to ambient level. The cool seed is stored.
6. **Bag drying:** The threshed and cleaned seeds are stored in gunny bags of small sizes. These bags are placed in the drying bin directly for drying seeds by hot air. It is an advantageous method to dry the small seeds of lots of many crops/varieties continuously.

Recommended temperature and depth for heated air drying of crops in bin

Crop	Depth in cm	Drying temperature (°C)	Safe moisture %
Maize	125	40-45	13
Wheat	1251	40-45	12
Sorghum	125	40-45	12
Soyabean	125	30-45	11
Rice	112.3	40-45	12
Groundnut	287.3	36	

7. **Solar dryers:** It is an improved technology for drying seed by heating air with the help of solar energy in a solar collector and then passed through beds of seed by a fan.

8. **Solar chimney:** Solar chimney provides a column of air that increases the flow of warm air in the dryer.

Ultra drying of seed for long term storage: The longevity of seeds can be increased by 4–16 times by drying them down to a moisture content of 1%–3% (Harrington, 1972). For long-term preservation, ultra-dried seeds don't need to be kept at extremely low temperatures. It is a useful technique for keeping conventional seeds stored. One way to do that is by drying silica gel. The best material for absorbing moisture and ultra-drying of seeds is silica gel. A bead with a diameter of roughly 1/16" to 1/8" works best for drying seeds. Up to 20% of the weight of dehydrated silica gel can be absorbed by it in water. It adjusts to the restricted atmosphere until the relative humidity in the controlled environment is between 10 and 12%. A little amount of cobalt chloride treated silica gel functions as a moisture indicator.

Lyophilization (freeze drying): By first freezing the seed water and then sublimating the ice, the seed is ultra-dried. The vitality of freeze-dried seeds is not affected in any way.

Management of drying operations: The seeds are dried immediately after receipt. In case of delay, it is aerated by fan to prevent heating of seed. The temperature and moisture content of seeds at different drying zone should be monitored regularly. Air flow, time of drying, static pressure, relative humidity and temperature of air, depth of seed etc. should be monitored regularly. The air forced in for drying of seed should be too hot as it may deteriorate the seed present at the bottom. During the process of drying, the hot air gets cool as it rises through seed lot and its capacity to hold water decreases due to saturation. The air deposits water on the surface of the seed present at the top layer. That results in transfer of water from bottom to top layer. Therefore, in the early stage of drying, when seed hold maximum amount of water, natural air should be forced into, and the temperature should be raised with advancement of drying period.

Tempering period: For minimum loss in viability of seed of maize cooling aeration is provided after high temperature air drying. Less than 1% moisture is removed when cooling is carried out immediately after drying. Whereas more than 2% moisture is removed when tempering period is provided after high temperature and slow cooling. Tempering phase enhances drying efficiency and reduces damage to the seed.

Two stage seed drying: During peak period, two stage seed drying may be adopted when limited drying resources are available. In the first stage of drying, the seed is immediately dried up to intermediate moisture content and stored. In the second phase, the seed is again dried after 20-30 days up to desirable level of moisture content. The method is adopted for rice and maize as rice seed can be stored up to 20 days at 18% and maize at 20% moisture content without significant losses either in quantity or quality of seed. After 20 days, the seed is dried properly for long term storage.

Pre drying aeration: Freshly harvested seed of rice with high moisture content can be stored safely up to 3-7 days with aeration of ambient air for 6-8 hrs. each day. The duration up to which rice seeds can be stored safely depends on the moisture content of seeds and ambient air conditions.

Drying of seeds: Seed should be dried at low temperature under high air flow rates than recommended for grain. Dryers and associated equipment should be designed for easy and proper cleaning to avoid mechanical mixture. For drying of seed continuous flow dryers are not recommended.

Prediction of Drying Period

The drying period is predicted by following methods:

1. Loss of weight after drying: The time required to dry the seeds until the equilibrium moisture content is predicted by loss of weight in the seed before and after drying.

a. Initial weight of seed sample before drying

b. The moisture content of the seed before drying is measured by oven dry method.

c. The percentage moisture content required for safe storage of seeds.

The process of the drying is started with weighing of seed sample at regular intervals until the weight of the seed reaches to the calculated value.

Weight of seed for safe storage

= Initial weight of seed × (100- initial moisture (%) of seed before drying

(100-Required moisture (%) of seed for safe storage.

Example

Initial weight of the seed sample before drying -100g

Moisture content of the seed before drying-14%

Moisture content required for safe storage-8%

What will be the weight of the seed sample after drying?

$$\text{Weight of seed for safe storage} = 100\text{g} \times \frac{(100-14\%)}{(100-8\%)}$$

$$\text{Weight of seed for safe storage} = 100\text{g} \times \frac{86}{92}$$

Weight of seed for safe storage = 100g × 0.9347

Weight of seed for safe storage = 93.47g

It shows that the moisture content of the seed will reach up to 8% when the seed samples of 100g of 14% moisture content dried to the weight of 93.47g. The time required for reduction in seed weight up to 93.47 g will be the drying period.

Mean Drying Curve

- Initial moisture content of the seed is determined.
- A large sample of seed is placed for drying as per the required procedure.
- The moisture content of the sample is determined each day till it reduces up to desirable level.
- A graph of drying curve i.e., mean percentage moisture content against day is plotted.

References

Khare, D and Bhale, M.S (2014). Seed Technology, 2nd revised and enlarged edition. Scientific publishers India, 5A, New Pali Road, P.O Box 91, Jodhpur-342001

7

Seed Delinting Methods and Procedures and its Assessment of Mechanical Damage

Priyanka Sharma

Concept of Delinting

Cotton seed clumps because of linters getting trapped on them after ginning. Gin-run seeds have very low flowability and do not singulate, making cleaning, upgrading, and precise metering in planting operations challenging or impossible. To singulate the seed and increase flowability, a variety of techniques have been and are employed. Most of the techniques entail removing the tags and linters entirely or partially, although other coating techniques have also been attempted with mixed results, both financially and technically (Mezynski, 1966; Webber and Boykin, 1907).

Methods of Delinting

1. **Mechanical Delinting**

Mechanical delinting is the usual technique used to improve the flowability of cotton seed. As previously indicated, mechanical delinting basically involves resharpening with finer, closer-spaced saws to remove part of the commercially valuable linters.

Although mechanical delinting increases the seed's flowability, it is insufficient to enable the precise conditioning procedures needed to distinguish between despined cockleburs and low-density, immature seed (Bunch *et al.*, 1961; Mezynski, 1966). Plant ability is also increased, although less precise measuring is required than with smoothly flowing seed. Mechanical damage is the main impact of mechanical delinting on seed quality, aside from enhanced flowability. The development of additional or alternative techniques for the partial or total removal of the linters resulted from the limitations of mechanical delinting with respect to flowability improvements.

2. **Flame Delinting**

To burn off loose linters, mechanically delinted seeds are dropped through a strong flame in a process known as flame delinting, or "flame zipping." Although flowability has significantly increased, it is still insufficient for activities involving precise cleaning and conditioning. Rapid "de-sparking" and seed cooling are essential because the seeds are heated as they pass through the flame and are touched by the burning linters. Heat can seriously harm a seed if these actions are not completed quickly and efficiently.

3. **Acid Delinting**

Wet-acid, gas-acid, and diluted wet-acid are the three primary types of acid delinting systems in use, according to Jones (1980). While the last method yields lint-free "black" seed or partially but uniformly delivered seed, the first two systems generate lint-free seed with outstanding flowability. Inadequate control and management of any acid delinting system might result in a reduction of seed quality.

Gas-acid Process

The gas-acid delinting method is mostly employed in dry regions where the moisture content of cotton seed is less than 9% and low humidity levels prevent facility and equipment corrosion. The linters are broken down by anhydrous HCl gas such that frictional forces can remove them from the seed (Jones *et al.*, 1974). Figure 1 depicts a generalised schematic of the gas-acid reaction. The seeds are rough cleaned to get rid of any large impurities after being dried as needed to bring the moisture level down to 5–7%. The gas-acid is then injected at a concentration of 0.5-2.0 percent of seed weight after a charge of seed has been placed in a revolving reaction chamber and the temperature has been raised to 60–70 °C. Depending on the temperature, seed moisture level, gas-acid concentration, and variety, the reaction time can range from five to twenty minutes. Following their escape from the reaction chamber, the seed goes via a reel where the degraded linters are completely removed by frictional forces. Ammonia is typically used to perform neutralisation. After that, the easily flowable, lint-free seed can be cleaned, graded, treated, and packaged much like other types of seed. To effectively delint without harming the seed, the gas-acid delinting process necessitates highly complex equipment, close supervision, and strict control over the numerous procedures. The main factors that harm seeds include excessively high reaction temperatures and gas-acid concentrations, as well as "over" neutralisation with ammonia and lengthy reaction times. Deficient management and control of gas-acid delinting might result in a significant decrease in germination and vitality.

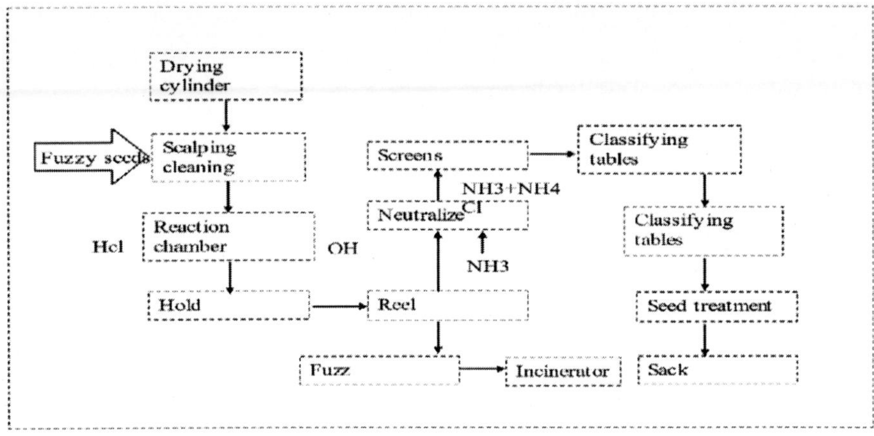

Fig. 1: Flowchart of cotton gas acid delinting of cotton seed (Source: modified from Jones, 1980)

Wet acid Process-

Throughout the cotton belt, the wet acid delinting method is preferred in humid, rainy regions. It is not necessary to use sophisticated equipment for this very basic process (Figure 2). Concentrated sulfuric acid is combined with gin-run seed and placed into a reactor trough or tank. The seeds are run through washers after being placed in the reaction trough, where the remaining acid and broken linters are removed. The seeds must be dried since they are moist before proceeding to the line for cleaning, grading, treatment, and packaging. When reaction times are longer than necessary, temperatures climb too high during the drying phase, the seed delinted has low vigour, and mechanical damage occurs more frequently than 12–15 percent of the time, seed quality losses can occur in the wet-acid delinting process. Aside from the expensive expense of sulfuric acid, the main issue with wet acid delinting is how to dispose of the spent acid and wash water. Previously, the wastewater was typically disposed of in a stream. Regulations and environmental concerns have eliminated this simple option (Sigman, 1973). There are environmental issues with the alternate approach, which involves collecting the sewage in a lagoon like a sewer.

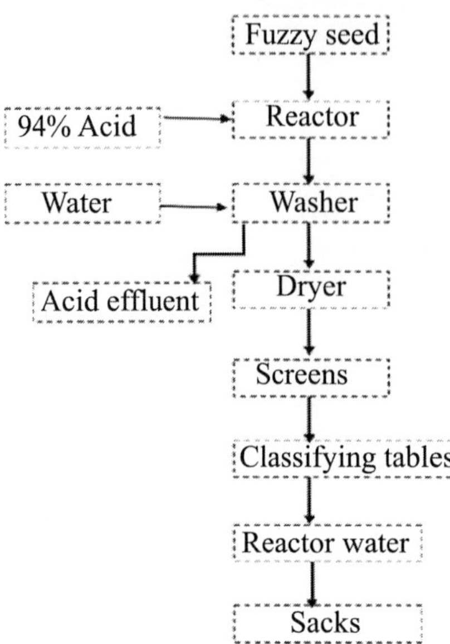

Fig. 2: Flowchart for conventional wet acid delinting process

Dilute Wet-acid Process

Cotton, Inc. invented the diluted wet acid delinting method (Jones, 1980; Jones and Slater, 1976). Figure 3 illustrates how this process varies from the traditional wet-acid method: the linters are wetted with a diluted solution of sulfuric acid (roughly 10 percent) rather than concentrated sulfuric acid; the wet seeds are "dewatered" by centrifugation to a level of approximately 10 percent add-on of the diluted acid; the seeds are dried with heated air to evaporate water, increasing the concentration of the acid; the degraded linters are removed by frictional forces in a rotating buffer-drum; the residual acidity is neutralized by adding ammonia or adding lime during the seed treatment procedure. The benefits of the diluted wet acid delinting method include recovering a significant amount of the sulfuric acid used in dewatering and eliminating the effluent created in the process, which results in a significant reduction in the amount of sulfuric acid required. The hydrolysed linters that are removed during buffing may also be useful for adding to animal feed and producing ethanol.

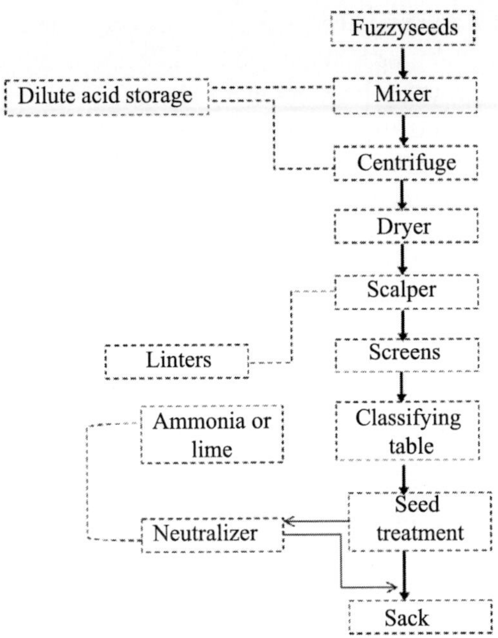

Fig. 3: Flowchart for dilute sulphuric acid delinting of cotton seed (modified from Jones, 1980)

Consequences of Mechanical Damage

Seed damage caused by mechanical means can have both immediate and latent effects. It may also have unintended consequences. When there is significant injury or damage to a sensitive area, such the radicle, the capacity to germinate may be lost immediately (Atkin, 1957; Keith, 1972; Klein and Harmond, 1966; Toole and Toole, 1951). Less severe injury reduces storage life, vigor, and field emergence potential (Col wick*et al.*, 1972; Koehler, 1957) and causes seedling abnormalities (Spreafico, 1965). When mechanical damage occurs, its indirect repercussions are frequently just as significant as its direct ones. Seed rotting bacteria, which easily enter necrotic tissue through cracks or cuts in the seed coat, are more likely to cause damage to damaged seeds in the soil (Erwin *et al.*, 1964). Additionally, procedures and materials employed in the processing of seed for commercialization are more likely to affect mechanically damaged seeds. Cuts in the seed coat during cotton seed acid delinting expose the embryo to acid, resulting in acid burn (Col wick *et al.*, 1972). Chemical seed treatments, including the once-common organic mercurial (Roane and Starling, 1958) and some of the systemic insecticides sprayed on cotton seed (Colwick *et al.*, 1972), frequently cause damage to seed. The effects of mechanical damage on cotton seed quality show that as the frequency and severity of mechanical damage increase, germination and storability decrease. Necrosis begins in the embryonic tissue beneath cuts and fractures in damaged seed. While treating damaged seed with fungicides improves laboratory germination, cold test, and field emergence, the negative effects of acid delinting (the conventional wet-acid process) worsen with increasing mechanical injury incidence. One in three severely damaged seeds, one in four somewhat damaged seeds, and one in five immature seeds in commercially processed seed lots are viable. Gin-run seed may be promptly tested for immaturity and empty seed using X-ray analysis.

Reasons for Affecting Seed Quality

Decreases in seed quality during the different processes are typically linked to mechanical harm, chemical damage, or physiological degradation brought on by high temperatures, high moisture content, and their interactions. Effective equipment selection and modification, improved facility design, enhanced operational management, and a strict quality assurance programme can all help to reduce these quality losses.

References

Atkin, J. C., (1957). Bean seed injury and germination. N.Y. Agr. Exp. Sta: Farm Research. April 1957. pp. 10-1.

Bunch, H. D., J. J. Da le, Jr., and W. W. Guerry. (1961). Separation of from acid-delinted cotton seed. Miss. Agr. Exp. St a. Inf. Sheet No. 714.

Col wick, R. F., T. H. Garner, G. D. Christenbury, G. B. Welch, R. L. Clark, J. C. De louche, C. C. Bas ki n, J. W. Sorenson, L. H. Wilkes, N. K. Person, and H. W. Schroeder. 1972. Factors affecting cotton seed damage in harvesting and handling. Agricultural Research Service, USDA. Prod. Res. Rept. No. 1.

Col wick, R. F., T. H. Garner, G. D. Christenbury, G. B. Welch, R. L. Clark, J. C. De louche, C. C. Bas ki n, J. W. Sorenson, L. H. Wilkes, N. K. Person, and H. W. Schroeder. 1972. Factors affecting cotton seed damage in harvesting and handling. Agricultura l Research Service, USDA. Prod. Res. Rept. No. 135.

Erwin, D. C., W. H. Isom, and M. J. Garber. 1964. Flax seedling emergence as affected by harvester injury, fungicidal treatments and age of seed. Crop Sci. 4: 217-219.

Jones, J. K., (1974). Gas-acid delinting of cotton planting seed. Cotton Inc., Memphis, TN. Special Research Report. April 1974. 15 pp.

Jones, J. K., and G. A. Slater. (1976). Di lute sulfuric acid cotton planting seed delinting process. Cotton, Inc., Memphis, TN. Agro-Industrial Rept. 3(1): 1-26.

Jones, J. K. 1980. Acid delinting cottonseed for oil milling and cubic acid delinted hulls. Oil Mill Gazetter. Sept. 1980. pp. 11-22.

Keith, G. (1972). The challenge in producing seed quality soybeans. In: Gra in Damage Symposium, D. Byg, Chm. Ohio State University, Columbus, Ohio. pp. 53-57.

Klein, L. M., and J. E. Harmond. (1966). Effect of cylinder speed and clearance on threshing cylinders in combining crimson clover. Trans. of the ASAE 9: 499-500, 506.

Koehler, B. 1957. Pericarp injuries in seed corn. Ill. Agr. Exp. Sta. Bull. 617.

Me zynski, P. R. (1966). Mechanical and electrical separation of despined cockle burs from mechanically delinted cotton seed. Ph. D. Dissertation. Missisippie State University, Miss. State, MS. 81 pp.

Roane, C. W., and T. M. Starling. 1958. Effects of mercury fungicides and insecticides on germination, stand and yield of sound and damaged seed wheat. Phyt opath. 48: 21 9-233.

Sigman, J. C., (1973). Char act erization and treat- ability studi es of acid cottonseed delinting waste. M.S. Thesis. Mississippi State University, Miss. State, MS. 88 pp.

Spreafico, L. (1965). normal seedlings in wheat. Injured seed, germination and abnormal seedlings in wheat. Proc. Int. Seed Test. Assoc. 30: 727-736.

Toole, E. H., and V. K. Toole. (1951). Injury to seed beans during threshing and processing. USDA. Circular No. 874.

Webber, H. J., and E. B. Boykin, (1907). The advantage of planting heavy cotton seed. USDA Farmers' Bulletin 285. 16 pp.

8

Seed Packaging, Selection and Effect of Packaging Materials on Seed Longevity and Handling

Meghali Barua and Aradhana Phukan

During processing seeds are cleaned, dried, and graded which are then treated with specific fungicide and insecticide. Seed at this stage is at best quality, which are then packaged into various packaging materials in a specific weight. Packaging or bagging of processed and treated seed is a requisite for easy handling of the bulk seed. Seed packaging can be defined as filling with seed, weighing and sealing the container whereas seed package as storage container where the seeds are kept for marketing or shipping or till the next sowing time.

The Packaging of Seed Involves

1. Filling bags/ containers of uniform size with seeds of known weight.
2. Providing leaflets inside the package with information about any specific cultural practices needed.
3. Attaching labels or certification tags having information regarding all the quality parameters.
4. Temper-proof sealing of the bag.
5. Storage or transportation of the bags.

Objective of Packing Seeds are

1. For easy identification.
2. To prevent seeds from absorbing moisture.
3. To avoid mechanical mixture with other seeds.
4. To protect seeds from insect pests, contamination, mechanical damage.
5. To prevent spillage and loss.

6. To conserve seed for long duration
7. For easy handling and transportation.
8. To increase the market value by enhancing attractiveness.

Selection of Packaging Material Depends on Various Factors

1. Seed kind
2. Seed quantity
3. Seed moisture content
4. Seed value in the market
5. Packaging material cost
6. Storage condition and environment
7. Storage duration
8. Transportation mode and distance

Packaging Method

1. **Area of packaging** should have lower relative humidity or seed should be exposed to outside environment for minimum time possible otherwise seed may absorb moisture. Dehumidifier can be used to minimize the RH of the packaging area.
2. **Selection of packaging material**: The material used for packaging is of paramount factor for maintaining the seed quality for longer time. Seeds are packed primarily for immediate or for future marketing; to keep seed for next season sowing; and for long term conservation.

Since seed always try to maintain an equilibrium between its moisture content and the relative humidity of the environment, hence, if not packed properly the dry seed would absorb moisture from the outside environment. Additionally, inappropriate packaging would increase possibility of leakage of seed or chemicals that may have been added to it, or entry of insect pest and pathogen. It will be more damaging if seed is shipped for long distance in a various mode of transportation. For this reason, packaging material should be selected judiciously.

Packaging Materials Should have Following Properties

i. Should be able to protect the seed quality.

ii. Should have sufficient tensile and bursting strength and tearing resistance to withstand the handling stresses.

iii. Should protect the seeds from insect pests.

iv. Should not allow seed to increase moisture content.

v. Should be easy to handle.

Packaging Materials can be Classified as

i. **Moisture and vapor permeable containers**: These containers allow free transmission of moisture and gases into the packaging materials from outside environment and vice versa, both in the form of vapor and liquid, hence suited for short term or temporary storage only. These are cheaper, can be reused, environmentally safe, and sufficiently protect the seed from spillage and mechanical mixture. Nonetheless the possibility of insect-pest, disease causing pathogens, rodents attack increases. e.g., cloth bags, gunny bags, paper bags, multiwall paper bag etc.

These packaging materials are mostly used when seeds are not dried to correct moisture content. Seed to be stored in bulk for one season in an ambient condition e.g., paddy, ground nut etc. could be packed in gunny bags, rigid plastic bags etc. Mustard and rape seed remain sufficiently viable and healthy for the next season sowing in a cloth bags. Seeds needing further drying or processing, or to be stored for short duration can be packed in such packing materials.

Recalcitrant seed, which cannot tolerate desiccation, should be stored in permeable bags for free exchange of air to avoid overheating due to high respiration, also to keep seed moist, periodic water spray can be done.

ii. **Moisture impervious but vapor pervious containers** i.e., **moisture vapor resistant or semi-impervious containers**: These are resistant to moisture but over a time there will be equilibrium between the relative humidity inside of the containers and outside environment due to entry of water vapor. Hence, it can be said that these allow entry of moisture into the packaging materials, in the form of vapor but not liquid. These also allow free exchange of gases. The seeds in the containers cannot be kept for very long period especially in hot humid conditions but can be kept longer than moisture and vapor pervious containers, though if seed moisture lowered to sufficiently safe level, the storage life can be extended. e.g., polythene bags of less than 700-gauge thickness, urea bags, polythene laminated jute/ cloth bag, aluminium foil, HDPE (high-density polyethylene) bags etc.

Seeds that are packed for further processing or for immediate use can be packed in simple and cheaper plastic zip-locked bags. Also, these can be used for packing seeds to be stored for short or medium duration storage. To avoid moisture-built up in such packages, desiccants e.g., silica gel, can be added with the seed.

iii. **Moisture and vapor proof containers**: These containers will not allow entry of moisture in any form. These are used for long term storage even in hot and humid conditions though moisture content of the seed should be optimum (safe level) at the time of sealing. e.g., polyethylene bags of more than 700-gauge thickness, laminated aluminium foil pouches, aluminium tins, rigid plastics, glass bottles etc.

For long term conservation e.g., the base collection of seed conservation where small number of seeds is stored for long duration, moisture and vapor proof containers such as glass vials, laminated aluminium foil pouches etc. are used for packing, sealed hermetically (airtight). (Base collection is stored for 50 or more years without losing viability. Though seed moisture content should be kept at very low (3-5%), storage temperature should be about -18^{0}C).

Vegetable seed which is a high value, but low volume seeds could be packed in rigid plastics pouches or metal containers with tight fitting lid sealed hermetically, of various sizes ranging from 50 g to 1 kg packets or containers with some variations.

During long distance transportation to protect seed from mechanical damage, high humidity and temperature, the seeds should be packed in moisture and vapor proof containers to prevent moisture absorption by the seed, more importantly when seed transit through humid region.

Though these are not suitable for recalcitrant seeds and seeds not dried to safe level. Additionally, these packaging materials are costly.

While packing seeds in any of the above-mentioned packaging containers especially in air-tight packages, lower seed moisture content would prolong the keeping quality of the seed. In India, under seed certification, the moisture content of the seed to be stored in vapour proof containers are kept at lower than the seeds to be kept in ordinary containers (Appendix-I). Higher seed moisture cause higher respiration rate, also induces fungal growth, ultimately causing seed deterioration.

For better tensile strength and protection from moisture and gases, two or more layers of packaging material can be combined. For fragile seeds, rigid-walled containers should be used.

IRRI Super Bags: These bags developed by IRRI is based on principle of hermetic storage but available at much lower cost than other similar packaging materials. It can be used to store cereal grains, crops like maize or coffee (difficult to store) for extended period from 9 to 12 months, to control insect pest without using chemicals because of reduction of oxygen and moisture exchange between the stored seed and environment outside, also there is further decrease in oxygen level inside (from 21% to 5%) due to respiration by grain and insects, which again reduces insects' population in the bag within 10 days of sealing. In rice use of these bags reduces grain cracking and increases rice recovery on milling. Up to 50 kg seeds can be stored in the bag.

Method of use: They are used as a liner inside other storage bags such as jute bags, HDPE bags etc. The bag is then filled with properly dried seeds but should never be over filled. The excess air is removed, and the bag is closed by twisting the free plastic portion above the grain, then it is folded and tied. Then the outer bag is closed. The damaged or puncture bag should not be used, also seed should always be carried by holding outer bag not the Super bag. Proper sealing, periodic checking should be done.

3. **Sealing**: After filling the packets or bags with seed, it should be sealed immediately by commercial sealers. Before sealing leaflets can be provided inside the package with information about any specific cultural practices needed. Cotton bags or gunny bags can be tied or sewed. Polyethylene or plastic bags can be sealed by heat sealers. Impermeable packages should be hermetically (airtight) sealed. Airtight containers not only prevent exchange of water vapor but also oxygen, which is a major cause of higher respiration rate, that leads to rapid seed deterioration. Glass jars or vials, metal cans, and laminated foil packets are considered best for airtight sealing.

Quality test of containers: There is a simple test to check sealing of containers.

i. Seal the container after filling with silica gel.

ii. Take the precise weight of the container.

iii. Keep it in a desiccator, filled with water, in the lower chamber for a week.

iv. Precautions should be taken to prevent contact between the container and water.

v. After one week, take out the container, remove all traces of water if present from container surface.

vi. Take weight.

vii. Any change in the weight will be indicative of improper sealing of the container, hence adequate measures should be taken such as increase in sealing time, replacement of container etc.

viii. After taking necessary measures, the test should be repeated.

4. **Labelling and tagging**: In India, the Seed Act, 1966 made the labelling compulsory, hence, any seed that are available for marketing or conservation should carry a label having all the requisite information. Seeds produced under seed certification must carry white and azure blue tag for foundation and certified seed respectively. Whereas breeder seeds carry golden colour tag. Truthfully labeled seed's tag is of opal green in colour. Information on the labels should be clear and in printed form. Packets or bags can be pre-printed with information regarding the name of the enterprise and logo (if any), net weight and precautions to be taken (if any).

For conservation, the labels of seed sample received, should carry following passport data-

A. Samples from collecting missions

i. Common crop name and/or genus and species

ii. Collecting number

iii. Location of collecting site

iv. Country of origin

v. Collecting date

vi. Phenology

vii. Collecting source

viii. Number of plants sampled.

B. Samples received as donations:

i. Common crop name and/or genus and species

ii. Accession name and/or other identification associated with the sample.

iii. Pedigree information and breeding institute's details (for breeding lines)

iv. Phenology

v. Acquisition source

vi. Country of origin

Certification tag along with net weight, pure seed (%), inert matter (%), germination (%),oil content (%) details, should have following contents and layout

<table>
<tr><td>TAG No.......................
KIND..........................
Variety.........................
Lot No..........................</td><td>CA's
EMBLEM
Name & Address of
Certification
Agency</td><td>Certified Seed
Class of seed...................
Certificate No.................
Date of issue of
Certificate.....................
Date of test....................</td></tr>
<tr><td colspan="2">"Use of the seed after expiry of the validity period by any person is entirely at his risk and the holder of the certificate shall not be responsible for any damage to the buyer of seed. No one should purchase the seed if seal or the certification tag has been tempered with"</td><td rowspan="2">Certificate valid up to.........................
(Provided seed is stored under cool and dry environment)
Validity of certificate further extended up to.................</td></tr>
<tr><td colspan="2">Name and Full Address of the Certified Seed Producer.......................................</td></tr>
</table>

(***Source:*** *Indian Minimum Seed Certification Standards, 2013*)

For seeds of Genetically Modified Crops, the labels should clearly state as '**Genetically modified**' and should have following details

i. Character under modification.

ii. Safety requirements or precautions to be taken for handling or storage.

iii. Contact information etc.

iv. Net weight

v. Date of expiry

For chemical treated seed, in the packages

i. It must mention as 'POISON' or with symbol which can be understood even by persons with reading difficulties.

ii. Name of the chemical used.

iii. Precautions to be taken for use.

iv. Instructions for accidental use.

v. Should clearly state 'Not for use as food or feed.'

a. Operate only by trained and skilled person.
b. Use thread of proper size
c. Sewing speed should be equal to the speed at which the sewing foot feeds the bag. Sewing the bag too rapidly or too slowly causes trouble.
d. Started into the bag on the side opposite the bag seam, to prevent jamming.
e. Operate with the proper thread tension to insure the correct stitch.
f. Operate with a smooth and properly adjusted looper. Knots of thread should never be removed from the looper with a sharp pointed or sharp-edged instrument; this may scratch the looper and cause thread cutting.
g. Maintain with well-oiled and clean. At frequent intervals the entire neck of the bag sewing should be dipped into a 50-50 mixture of kerosene and light motor oil and run for a few seconds as it removes dust and trash.
h. Prevented from striking a seed as the bag is sewn shut.

Seed Handling

Several types of conveyors are available for handling of unpacked or packed seeds in processing plants such as bucket elevators, belt conveyors, vibrating conveyors, pneumatic conveyors, screw conveyors, chain conveyors and lift trucks.

Bucket Elevators

Bucket elevators are widely used in agricultural industries including seed processing plants lifting seeds either vertically or in inclined position. It consists of series of buckets connected to an endless belt or chain. The belt or chain will be moving and hence bucket will also move upward. The spacing of the bucket and speed of the conveyor control the flow rate of seeds. As the bucket moves upward, bottom bucket occupies the position of feeding point. Thus, the movement continues. The material discharge could be by centrifugal force or by simply dumping tilting buckets. The assembly is usually enclosed in a steel or wooden casing.

Belt Conveyor

Belt conveyor is an endless belt which is held under tensions between two rollers one of which is driven. The belts may be stainless steel mesh, rubber or composite material made by canvas, polyethene or polyester. In some designs belt moves on roller for easy conveying. Belt conveyors are universal transportation equipment to carry heavy loads as high as 5000 ton per hour (tph) and belt travels with speed of 5 m/s. These are useful for pre cleaned seed, cleaned seed and bagged seed also.

Vibrating Conveyor

Vibrating conveyors can also be called as shakers. They are used for moving seed materials in a horizontal way or near horizontal angles. The conveyor is consisting of a trough supported by springs. Seeds fed into the trough move up and forward with each vibration. These conveyors are suitable for short distance horizontal conveying of seed inside a cleaning plant.

Pneumatic Conveyor

Pneumatic conveying is an excellent method for transportation of granular materials which are easily flow able. This conveying system uses a gas, usually air to transport the material. It consists of flexible system pipes through which small particles are suspended in recirculated air and transported. The critical air velocity in this type of conveyor carries some major disadvantages. If the velocity is low, the materials block the pipe and if it is too high, there is risk damage to internal pipe surface.

Screw Conveyor

Screw conveyors are used for transportation of bulk materials over long or short distance. Preferably they are used for short distance within the process plant or for feeding of raw material into hopper at predetermined rate. They consist of trough or long channel into which the screws in the form of helix rotate by means of electric motor. Generally, screws rotate with low rpm. The material is fed at one end. As the screw rotates, the helix pushes forward the material. They are not recommended for easily damaged seeds because of friction among seeds which may cause cracking.

Chain Conveyor

Chain conveyors are slow moving conveyor that require high power requirement. Chains are engaged to wheels and sprockets attached to head (front) and tail (opposite side of the front). The materials are carried when chain moves forward by rotating the wheels and sprockets. The conveyor works horizontally or at an inclination angle of 45°.

Lift Trucks

Lift trucks are well constructed boxes serve both as conveyor and storage. They are useful for handling large number of lots on relatively few processing machines. The boxes are equipped with movable tops and perforated bottoms, where temporary drying and storage can be done.

Conclusion

Seeds packed at appropriate time with suitable packaging material would prolong the seed life. Suitable packaging method will maintain the physical and physiological quality of the seed hence, selection of packaging method and the packaging material could be considered as crucial step in the seed processing.

Appendix-I: Moisture content of few crop seeds for packaging

Crop	Moisture content	
	Vapour pervious container	Vapour-proof container
Cereals & millets		
Barley, Wheat, Triticale, Maize, Sorghum, Pearl millet, Barnyard millet, Common millet, Finger millet, Foxtail Millet	12	8
Paddy	13	8
Pulses		
Black gram, green gram, Bengal gram, Cowpea, Horse gram, Indian bean, Khesari, Lentil, Moth bean, Peas, Pigeon pea	9	8
French bean	9	7
Vegetables		
Tomato, brinjal, chilli, capsicum	8	6
Indian bean, Pea	9	8
Lady's finger	10	8
Cole crops	7	5
Cucurbits	7	6
Radish, turnip	6	5
Carrot, Celery	8	7
Amaranthus, asparagus, lettuce, onion, fenugreek, TPS	8	6
Spinach, beet,	9	8
Oil seed crop		
Soybean	12	7
Groundnut, sesame, niger	9	5

Crop	Moisture content	
	Vapour pervious container	Vapour-proof container
Indian Rapeseed & Mustard	8	Mustard=5; Rapeseed=7
Rocket Salad (*Taramira*)	8	5
Sunflower, safflower, linseed	9	7
Castor	8	5
Fibre crops		
Cotton	10	6
Jute	9	7
Forage Crops		
Berseem, Indian Clover (Senji), Lucerne	10	7
Dinanath grass, forage sorghum, sudan grass, guinea grass, marvel grass, setaria grass, teosinte	10	8
Gaur (cluster bean), Ricebean	9	8
Oats	12	8
Green manures: Dhaincha, Mesta, Sunnhemp	9	8
Spices: Ajawain, cumin, coriander, fennel	10	8
Flowers		
Aster	9	6
Carnation, chrysanthemum, Petunia, Snapdragon	8	6
Marigold	9	7

(***Source:*** *Indian Minimum Seed Certification Standards, 2013*)

References

Agrawal RL (2013). Seed Production of oil crops. In book: Seed Technology, Publ.: Oxford & IBH Publishing Co. Pvt. Ltd. pp: 159-182.

Anonymous (2019). Study of existing supply chain processes in the seed sector in India terms of references. Ministry of Agriculture & Farmers Welfare (Department of Agriculture, Cooperation & Farmers Welfare). https://seednet.gov.in/

Bar Code India (2023). How Seed Traceability is Revolutionizing the Agriculture Sector? -BCI (barcodeindia.com).https://www.barcodeindia.com/blog/how-seed-traceability-is-revolutionizing-the-agriculture-sector

Dadlani M, Gupta A, Sinha SN and Kavali R (2023). Seed Storage and Packaging. In book: Seed Science and Technology by M. Dadlani, D. K. Yadava (eds.). Springer Nature Singapore Pte Ltd. pp-133-152. https://doi.org/10.1007/978-981-19-5888-5_7pp: 247-248.

Desai BB (2004). Seeds Handbook- Processing and Storage. CRC Press.

Indian Minimum Seed Certification Standards (2013). The Central Seed Certification Board, Department of Agriculture & Co-operation, Ministry of Agriculture, Government of India, New Delhi.

IRRI Super Bag - IRRI Rice Knowledge Bank http://www.knowledgebank.irri.org/step-by-step-production/postharvest/storage/grain-storage-systems/hermetic-storage-systems/irri-super-bag, accessed on 20.01.2024

Justice OL and Bass LN (1978). Principles and practices of seed storage. U. S. Government Printing Office. pp: 142-147.

Kugbei S, Avungana M. Hugo W (2017). Training Toolkit. Module2: Seed Processing equipment, The Food and Agriculture Organization of the United Nations and Africa Seeds, Rome.

Misra MK (1914). Issued in furtherance of Cooperative Extension work, Acts of May 8 and June 30, in cooperation with the U.S. Department of Agriculture. Stanley R. Johnson, director, Cooperative Extension Service, Iowa State University of Science and Technology, Ames, Iowa.

Stein WI, Slabaugh PE and Plummer AP (1974). Harvesting, processing and storage of fruits and seeds. In Seeds of Woody Plants in the United States, Agriculture Handbook No. 450. For. Service, USDA, Washington D.C.

Willan RL (1985). Chapter 7 SEED STORAGE. A Guide to Forest Seed Handling (fao.org). https://www.fao.org/3/AD232E/AD232E07.htm#ch7

www.Agrimoon.com developed by TNAU, Tamil Nadu.

9

Principles and Methods of Seed Extraction Methods in Vegetable Crops

Priyanka Sharma

Objective

Vegetable seed separation requires specialized training. A little carelessness during the seed extraction process can seriously harm the plant's viability and vigor in addition to its outward look. Inadequate extraction methods can also result in *in-situ* germination. The seed can be separated by the following methods.

1. Manual Method

(a) Maceration e.g., watermelon,

(b) Crushing e.g., brinjal,

(c) Scraping e.g., cucumber

(d) Separated e.g., muskmelon,

(e) Scooping e.g., pumpkins and

(f) Extraction e.g., squashes.

2. Acid Method

The completely ripened, mature fruits are collected using this approach, and they are then ground into pulp. The pulp is removed out of a convenient-sized cement tub, wooden container, or plastic container, and commercial HCL is added. After a thorough mixing, the pulp and acid are stored for that amount of time. During this time, the acid's corrosiveness breaks down the mucilage that is attached to the seed and releases the pulp from it. After that, the seeds are properly cleaned with water four or five times to remove any remaining acid, as this could damage the seed's embryo. This approach allows for a quicker extraction of seeds. Seed are also bright in colour with good germinability and free from fungal assault. Regardless of variety, the various extraction techniques revealed that the acid procedure had a higher seed recovery percentage. After 30 minutes of soaking in 2.5% HCL, germination was at its peak. The concentration of hydrochloric acid (HCL) in vegetables varies; table 1 illustrates this.

Table 1: HCl concentrations and time taken for seed extraction in different vegetable crops

Sr. no.	Crop name	HCL concentration	Time taken
1	Tomato	25 ml/1 kg of pulp	30 minutes
2	Brinjal	4 ml/1 kg of pulp 10 ml/1 kg of pulp 30 ml/1 kg of pulp	60 minutes 45 minutes 20 minutes
3	Watermelon	1:6 (Acid: water ratio)	2 hours
4	Pumpkin	1:6 (Acid: water ratio)	5 minutes
5	Cucumber	100 cc/1 kg of pulp	30 minutes
6	Potato	10 ml/1 kg	20 minutes

3. Fermentation Method

After being crushed, the fruits are allowed to ferment for two to three days in a non-metallic container. It has been discovered that two days following fruit fermentation is the best time to harvest high-quality seed. During the fermenting process, the seeds break from the adhering pulp and settle to the bottom of the container. To get the appropriate moisture level, the seeds are separated, carefully cleaned, and dried in the shade. When compared to alternative extraction methods, the seed recovery is lower. Both the fungal load in the seeds and the pulp's fermentation cause the seeds to turn a drab tint. Extended fermentation times can lead to *in situ* germination. These are method used in vegetables like, tomato, brinjal, cucumber, watermelon, musk melon, etc.

4. Mechanical Seed Extraction

This technique is primarily applied to tomatoes, brinjal, and chillies. The pulper machine receives a known weight of ripe tomato fruits as input. After being removed individually from the outlet, the pulp containing the seed is cleaned with water and allowed to dry in the shade. Pulpers can also be used to crush the fruits of brinjal. Enough water is added before utilizing pulpers, and the pulp is thoroughly stirred after pulping. The maximum seed rate extraction (3.327 kg/hr) is achieved when utilizing a vegetable seed extractor and the treatment combination of 2 mm concave clearance + 8.5 m/s cylinder peripheral speed + 1.76 of hourly feed rate. Dried chilli fruits are put into the feed hopper of a seed extractor, where they are beaten. This separates the seeds, which are then released out the outlet. The seed are manually separated from the hulls. 96% of the seeds are extracted efficiently.

5. Alkali Method

Fruits that are fully grown and mature are gathered and mashed into pulp. Tomato pulp is treated with 0.5% sodium bicarbonate (500 g diluted in 10 l of warm water) and allowed to ferment for one day. After that, the seeds are divided and washed with water to get rid of any leftover alkali.

6. Citric Acid Method

This method of extracting seeds eliminates the gelatinous coating from seeds without compromising their germination or vigour. It involves adding 30 g of citric acid to one litre of pulp and allowing it to digest for two hours. However, this procedure was only shown to work with tomatoes and was proven to influence the seed's storability.

7. Modified acid Method

Freshly picked fruits are pulped using water. The wet seed with mucilage remains after the pulp and peals are removed. Ten kilos of fruit will produce one kg of moist seed. Add forty ml of commercial HCL to this and stir continuously for twenty minutes to allow the mixture to react. After that, the seeds are dried and cleaned. This technique preserves acidity without compromising seed quality or recovery. It's used in Tomato, Watermelon and Musk melon.

8. Dry Method

When the vegetables have fully grown, they are harvested and allowed to dry for two to three days in the sun. To restore the original moisture content, the seeds are dried in the sun between the hours of 8 a.m. and 11 a.m. and 2 p.m. after being extracted. Examples include ridge gourds, spinach gourds, okra, and chillies (Figure 1).

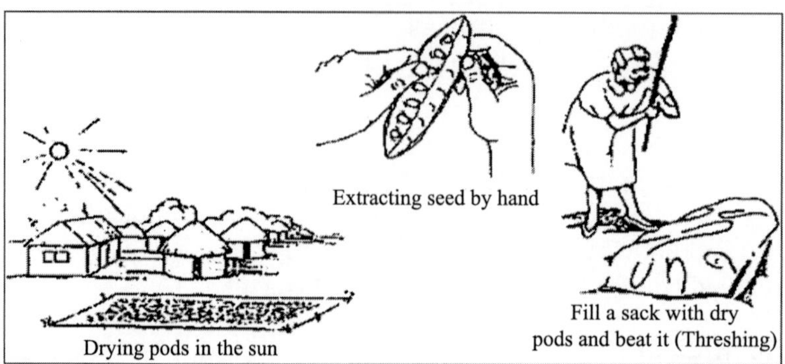

Fig. 1: Dry method of seed extraction in legumes

9. Directly Harvesting of Matured Pod Method

When the pods reach maturity, they are harvested and sun-dried for two to three days, bringing the moisture content down to 15–16%. The seeds are then extracted by beating the pods with a flexible bamboo stick. Reducing mechanical damage to the seed can be achieved by avoiding excessive drying and vigorous beating. Lower seed quality is the result of excessive mechanical damage. This process is primarily applied to cluster beans, cowpeas, French beans, okra, and so on.

10. Manually Seed Extraction Method

In this method, the fruits are manually seeded and chopped into longitudinal pieces. The seeds are dried, and the pulp remnants are cleaned. Seed storability is also significantly impacted by this technique. Examples include bitter gourd, pumpkin, watermelon, and Musk melon.

11. Floatation Method

Using the flotation technique, the sinkers and floaters are separated. As floaters, the immature seeds can be extracted. Bottle gourds and other large seed vegetables are two examples.

12. Wet Method

This technique was used using sweet pepper. After the fruits are smashed, the seeds are mechanically extracted from the remaining fruit pulp and debris. Typically, the crushed material is fed through a rotating cylindrical screen to separate the seed from the detritus.

Seed Curing of Vegetables

To prevent branches from drying out too quickly, the harvested crop is heaped up in small heaps and covered with hay or tarpaulin while cured on a cement floor. Curing with branches reduces shattering losses in the field, enhances seed colour, and allows immature seeds to ripen more slowly. The heap is flipped over and left to cure for an additional four to five days after the first four. When harvesting during a wet, humid season, caution should be exercised, and the stock should not be stored in this condition for longer than four to five days. This technique is applied on cauliflower and cabbage. Paddy or wheat straw is used to cover heaps to shield them from the sun. Tarpaulins should be placed over the mounds in case it rains. Compared to sandy loam, heavy treatment soil may need more time to cure.

Seed Drying of Vegetables

When seeds are harvested, their natural moisture content is frequently higher than what is needed for the optimal germination and longest possible life. The most significant element affecting the seed's lifetime and ability to germinate is most likely its moisture content. The moisture content of the seed depends on its relative humidity once it separates from the parent plant and is balanced with the surrounding humidity. Melons, tomatoes, and cucumbers are examples of fleshy fruits whose seeds have a higher moisture content upon harvest and may absorb more water during the wet extraction process. Conversely, seeds from fruits—like onions, brassicas, etc.—that desiccate during the ripening process are generally dry when they are harvested.

Freshly collected vegetable and flower seeds may have a moisture content of 18 to 35% in humid tropical environments; this moisture content needs to be lowered to a "safe level."

Cleaning of seeds

Some debris, such as pieces of cones, capsules, pods, etc., is also present with the seed when the fruit's seeds are extracted. There are also a few tainted seeds there as well. As a result, washing every piece of material is crucial to preventing germination issues.

Pre-cleaning

The condition of the seed must be assessed both after collection and prior to storage. When it comes to pods, it gets faster. To prevent further fermentation, moist and fomenting pods need to be fanned out to dry and treated with fungicide.

Methods of Cleaning

1. **Water Method:** Add the trash and seeds to the water. Debris will begin to float on the surface as the seeds settle. Clear away any rubbish and allow it to dry.
2. **Winnowing method:** Place the seeds in the device for winnowing with the detritus. Seeds will fall to the ground and debris will blow off if the material is dropped from a height towards the wind.
3. **Hand picking:** Clean the seeds with your hands if the debris that has been mixed in with them is easily removed.
4. **Sieving method:** When the debris is less in size and quantity than seed, this method can be used. This approach involves placing seeds in sieves with debris. The seeds stay in the site while the rubbish passes through the sieves due to shaking.

10

Principles of Seed Blending and its Methods and Practices in Seed Lots

Priyanka Sharma

Concept: Seed blending is the process of merging two or more types of the same species, for as Kentucky bluegrass cultivars Merion, Baron, and Park. Contrarily, seed mixing refers to the blending of two or more species, however every species may have more than one variation represented, such as Jamestown and Cascade chewing fescue combined with Merion and Fylking Kentucky bluegrass. Seed is typically distributed in blends, primarily Kentucky bluegrass blends, for sod growing purposes. For home lawns, parks, and other utility areas where upkeep levels would not be adequate to maintain pure Kentucky bluegrass turf, mixes of different grasses are typically employed. Naturally, when it comes to pure seed quality, no blend of grass seeds is superior to the individual ingredients. Every time a new blend needs to be made; extreme caution must be taken to ensure that the mixing equipment can be completely cleaned to prevent any contamination from unwanted species—bent grasses are perhaps the most challenging to deal with.

Principle: Blending of a seed lot of above minimum seed certification standard with the seed lot of below minimum standard in such a way, that the final seed lot maintains all the standards above minimum certification standards.

Pearson square: When two seed lots are mixed then the seed certification standard of the prepared lot will be of intermediate quality. To prepare a seed lot of required standard both the seed lots must be mixed in a particular proportion. Pearson square is a way to determine the proportions of different lots that can be blended to prepare a final lot with the desirable quality.

Example: Lot 'A' of wheat has 64% germination, whereas lot 'B' has 96%. The minimum seed certification standard for germination of wheat is 85%. If the required germination of blended lot is 86% then in what proportions these two lots can be blended to give a final lot?

Procedure: A square is drawn and the required germination percentage (86%) of blended lot is put at the centre. Germination percentage of the lot A and B are put on each of the left-hand corner of the square. Lines are drawn diagonally across the square and the value placed in the centre is subtracted by the values placed on left corners irrespective of larger figure.

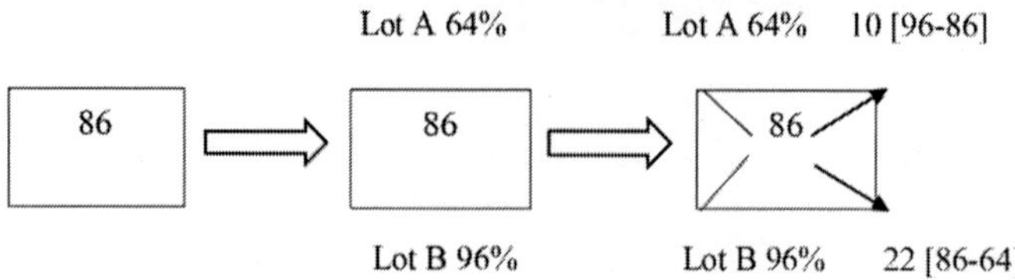

The results are written diagonally across the square from the relevant percentage on the left corner to the right-hand corners. The portion of the seed directly on the left corner, represented by the figure in each right-hand corner, can be combined to create the last lot of the germination percentage, which is then placed in the square's center. As a result, combining 10 parts from lot A and 22 parts from lot B will result in an 86% germination final lot.

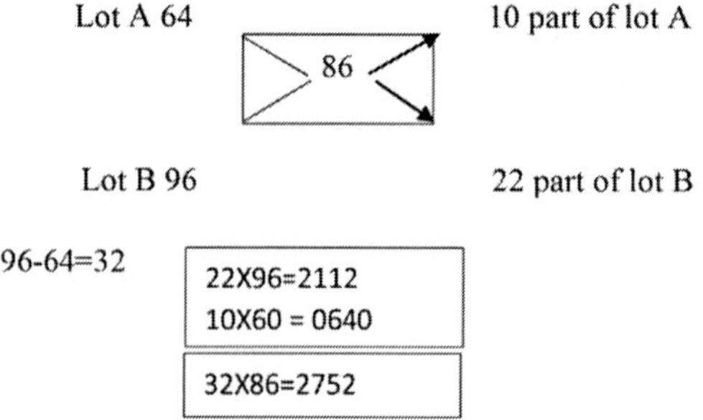

Seed lot averaging much below the seed certification standard and affected by disease or insects should be considered for blending. The Pearson square can also be used to determine the proportion for blending of more than two seed lots. The basic requirement for blending of more than two seed lots is the number of seed lots with seed standard higher and lower than the value of final seed lot should be same. In the event, the number is not equal than one lot of the category with less number is not equal than one lot of the category with less number must represent twice in the square.

Example: In what proportions the seed lots of soyabean with 88, 80, 77, 66, and 60 percent germination will be blended to produce one lot of 74% germination?

Write the required value of final seed lot at the centre of the square. Subtract the value diagonally i.e. 88-74 = 14 and write it on the opposite side at the lowest corner of the Pearson square. In this way, write all the values after subtraction. In the event germination percentage of the lot is more than the required germination percentage of the final lot, subtract larger value from the smaller value i.e., 88-74. When required percentage is more than the germination percentage of seed lot again subtract higher value with lower value i.e., 74-66 =08.

It shows that to prepare a final seed lot with 74% germination, 14 part of the seed from seed lot with 88%. 12 parts of the seed lot with 80%, 8 parts of the seed lot with 77%, 3 parts of the seed lot with 66%, 6 parts of the seed lot with 62%, and 14 parts of the seed lot with 60% germination should be blended. The proportion of the quantity to be blended of each seed lot may vary with the change of the position of the seed lot on the Pearson square with the same result.

References

Khare, D., Bhale, M.S (2014). Seed Technology (2nd) revised and enlarged edition. Scientific Publishers India, ISBN:978-81-72338-84-8. 5A, New Pali Road, P.O. Box 91, Jodhpur-342001, India.

Ledeboer, F.B., (2000). Seed Quality: Purity And Viability - Seed Blending And Mixing F. Agronomy and Research Loftis Pedigreed Seeds Inc. Bound Brook, New Jersey.

11

Pre-Sowing Seed Treatments for Seed Invigoration

Priyanka Sharma

Pre-Treatment Before Sowing

The natural conditions necessary for seeds to germinate are increased by pre-sowing seed treatment. Many plants and trees' seeds require a specific amount of time to mature in their immature portions and to overcome dormancy, which is the absence of certain elements that impede germination, so that the seeds can sprout. Depending on the type of seed, different environmental factors promote different levels of germination. For example, some seeds need to be kept in a very damp, cold environment, while others need to be kept in a dry, hot environment. Different techniques are employed to support seed survival and speed up germination, depending on the size of the lot, the species, the local conditions, and the availability of pertinent tools and facilities. Many species' seeds don't germinate successfully unless they are exposed to specific circumstances. Dormancy is the state in which seeds do not germinate unless certain circumstances are met. Conditions in the natural world could include being near fire or being eaten by animals. When seeds are eaten by animals, their stomachs' hydrochloric acid exposes them, breaking the dormancy of the seeds without causing harm to them. Man treats seeds in a similar way to induce germination by breaking their dormancy. Tree seeds can be pre-treated in a variety of ways, but practically all species can be successfully germinated if one knows a few basic approaches. As was previously noted, water imbibition and gaseous exchange are the two processes that must occur for Prosopis seed to germinate. However, the seeds may go dormant due to a variety of inhibitions brought on by the seed envelopes, such as permeability to oxygen, radicle protrusion, water, or mechanical barriers. Less than 5% of prosopis seeds with undamaged endocarps have germinated despite repeated tries. The seed needs to be scarified—that is, its seed coat must be scratched to promote germination—or otherwise treated to make it permeable to water and speed up germination to overcome its dormancy. The pre-treatments of seed prior to planting can be divided into five categories: electrical, chemical, mechanical, water, and dry heat treatments.

Mechanical Treatment

Small amounts of seeds can be scarified in an efficient manner by using sand paper to make a tiny scratch on each seed, cutting each seed with a knife, or covering the end of each seed that is opposite the radicle with sandpaper until the cotyledon is visible. Large operations might not be able to implement these treatments, though, because each seed must be treated separately. Mechanical scarification can be accomplished for large amounts of seed by rubbing the seeds over an abrasive slab or smashing the seeds with sand. These two methods have been proven to be effective and are both easy to use and reasonably priced. Further enhancing germination can be achieved by lightly contacting the hard seed coat that has been scarified by the procedures. Shaking the seeds twice a second for around fifteen minutes in a glass or metal container is the easiest approach to get a light impaction. By substituting an adhesive-backed sandpaper and solid sheet metal for the threshing screen, one can also achieve mechanical scarification with the thresher outlined by Flynt and Morton (1969). If scarification is not done correctly, it might shorten the seed's storage life and result in a variable percentage of seedlings that are injured. As a result, great caution needs to be taken.

Water Treatment

To break through seed dormancy, cover the seed with boiling water and let it soak while the water cools for a whole day. This method, can be useful for raising germination rates and imbibition. Breaking dormancy in hard seed coat seeds typically requires more than just soaking them in tap water at room temperature.

Chemical Treatment

It is possible to scarify small samples of seed lots by immersing them in pure ethyl alcohol for a duration of 12 hours. If the acid is to be soaked for a fixed amount of time, a concentrated (98 percent) sulfuric acid treatment is sometimes advised for big seed batches. Soaking periods typically range from 15 to 30 minutes. The general explanation for the increase in germination following sulfuric acid treatment is that oxidation softens the seed coat, increasing the permeability of water and air through it. After soaking in the acid, the seed needs to be carefully rinsed several times with lots of water.

Electrical Treatment

According to research by Nelson *et al.* (1978), germination rates were raised when *Prosopis juliflora* seeds were exposed to varying durations of radiofrequency (RF) dielectric heating by electromagnetic fields of 10 megahertz (MHz) and 39 MHz. However, more research will be required to show that treating seeds with RF can effectively break dormancy.

Stratification

Combining seeds with wet media—such as sand, peat moss, crushed perlite, vermiculite, composted bark, or sawdust—that has been specially selected to replicate natural environmental conditions and accelerate seed germination is the most popular stratification approach. The seed mixture can be stored in polyethylene bags or plastic boxes with tight-fitting lids that are properly sealed. The seed-media mixture-filled containers are stored at room temperature for moist-warm therapy and in a refrigerator for moist-cold therapy. Most natural tree seeds need to be exposed to freezing temperatures (to mimic winter colds) in a continental climate, like Armenia's, to germinate in the spring.

Method of Stratification

Put a water proof marker on clean bags or containers and write the species, the date of stratification, and the sowing date later. As an alternative, number the containers clearly and record the information for each numbered container in the computer file or seed preparation log. For each species of seed being prepared, prepare the required amount of moist medium (sand, peat moss, perlite, vermiculite, composted bark, or sawdust). Ensure that there is enough room available for storing the bags or containers at the ideal humidity, temperature, and lighting levels. Keep in mind that certain seeds should be stored in naturally light or dark environments.

Mixing and Stratification

Most cultivators advise combining equal parts dry seeds and wet medium (50:50 by volume, not weight). The medium's water content must be such that it feels moist to the touch without creating a water layer at the bag's (container's) bottom. Fungi will overgrow in a sealed container that has too much water in it. Many industrial facilities add fungicide preparations to the medium to avoid fungal damage to the seeds. Care must be taken to measure and add the appropriate type of fungicide to provide the optimum outcomes. Put the sealed containers in a stratification environment with a temperature range of 0 to 5°C. Because the seeds depend on their reserves of organic and chemical energy to remain alive, never let the seed-medium mixture remain in stratification for longer than the necessary amount of time. The seeds frequently germinate during the stratification stage because of a rise in temperature. This could lead to the stratification process ending early. Therefore, it's crucial to handle the delicate seeds carefully and maintain the right temperatures and moisture levels.

Scarification

Many plant seeds have tough, dry shells that, in the wild, must be exposed to warmth and moisture for an extended period to soften. Scarification refers to techniques used to disrupt the integrity of the seed's outer shell to facilitate water entry inside the shell.

Methods of Scarification

In a concrete mixer, seeds are combined in volumetric proportions with coarse sand. Manual scarification of small quantity of seeds can be accomplished with a nail file, fingernail clipper, or other mechanical equipment. The shell is completely punctured during manual scarification, but caution is exercised to prevent harming the seed's embryo.

Chemical Scarification

Depending on the species and solution concentration, seeds are soaked in an acid solution (sulphuric acid being the most frequent) for a few minutes to many hours. This method might be the most successful, but it requires careful attention to acid concentration and time, as well as extra caution due to the caustic nature of the acids and the need for personal protection equipment for the skin and eyes.

Soaking in Hot Water

After adding enough boiling (up to 100 °C) water to completely submerge the seeds in the container, the water is left to cool to room temperature. Every batch or container is labelled with the relevant information, as was previously described. The required information consists of:

1. Number (s) and location of container (s)
2. The species and the origin (vendor, batch, etc.) of seeds; scarification method if used.
3. Date of scarification or stratification, name (s) of the person (s) and their signatures for accountability; conditions of stratification (medium, temperature, lighting, proportion of the mixture, etc).

4. Date of expected maturity for sowing; results of periodic checks (frequency depending on the species).
5. Other information as appropriate.

Pre Sowing Treatment of Tree Seed Species

Depending on the level of dormancy and the pre-sowing treatment technique, seed species should be grouped.

1. Seeds with Light or no Dormancy

Stratification of these seeds is not necessary, but it will expedite germination. After soaking seeds in room-temperature water for a full day, strain them. After that, store them in containers or plastic bags in the refrigerator (2 to 5° C) for two to four weeks. Example: Pine, Fir, Birch, Platan, etc.

2. Seeds with Embryo Dormancy

Without a cooling time, these seeds will not germinate. After soaking seeds in running water for 48 hours, strain and combine them with perlite or sifted sand. Depending on the species, store them in a refrigerator for 12 to 16 weeks after placing them in plastic bags and loosely tying them.

Seeds of tree species which can be treated by this method includes

Beech-12 weeks

Maple-16 weeks

Walnut-16 weeks

Wild pear-16 weeks

Wild apple-16 weeks

Hazel-16 weeks

Sorbus-16 weeks

3. Seeds with Deep Dormancy

These seeds require both warm and cold stratification to germinate. After soaking seeds in running water for 48 hours, strain and combine them with perlite or sifted sand. Fill polythene bags with seeds, then loosely knot. For two weeks, store the bags at room temperature (20°C) and out of direct sunlight. Retain the seed's moisture, but not its wetness. Wet the seeds one or two times a week. Place the seed packets in the refrigerator for 32–37 weeks, depending on the species, following the warm stratification. For instance, Juniper is 34 weeks, Tilia is 37 weeks, and Ash is 32 weeks.

4. Seed with Seed Coat Dormancy

Before these seeds may sprout, they need to be "scarified." Cover the seeds with 100° C boiling water. After the water reaches room temperature, let the seeds soak in it. Next, remove the seeds from the water and sow them. E.g.: *Robinia pseudocacia.*

Stratification in boxes, trenches or under the snow: Stratification can be done beneath the snow, in boxes, or in trenches for bigger seed lots that are used for production.

Stratification in boxes: In wet, well-ventilated environments (like a cellar), tree and shrub seeds that do not germinate in the year of planting without a pre-sowing treatment can be stratified.

Wooden 100x30/40 cm boxes containing seeds combined with peat or sand are drilled with 0.5–1 cm ventilation hole in rows, spaced around 5 cm apart within each row and 10 cm between the rows. In a cellar, boxes are arranged on wooden planks that are 3–4 cm thick or on shelves. Typically, stratification takes place between 1 and 5°C in temperature. The boxes are set on top of ice or snow in the event of early sprouting. However, trenches that are warm, cold, can also use this stratification technique.

Stratification in trenches: Additionally, trenches that are warm, cold, might be used for stratification. For seeds whose pre-sowing dormancy period extends beyond four months, warm trenches are employed. Typically, a warm trench is excavated in a dry area that is 80 cm deep and 100 cm broad, with length added as needed. Wooden boards are positioned on supports 20–25 cm above the trench's bottom; ventilation tubes and reed or brushwood bundles, each 20–30 cm thick, are spaced every 1.5–2 meters. After that, a layer of peat or sand measuring 10 cm is applied. After that, the trench is sealed with boards, loaded with a mixture of seeds and suitable substrate (sand, peat), and topped with a layer of thatch 20 cm thick. A shovel is used to mix the seed mixture every ten days before to the start of the cold season. The thatch layer that is distributed over the boards can thicken up to 55–75 cm as the colder months approach, then thin down to 35–40 cm as the snow cover approaches. Seeds that go through a two to four months dormant stage are cold adapted (freeze through trenches). The ditch is 100 cm wide and 60 cm deep. In the trench, boards are positioned 20 cm above the bottom. A 30-35 cm thick covering of seeds combined with peat or sand covers the hardwood floor, while a 10-15 cm thick layer of thatch covers the latter. Summer trenches are excavated to a width of 50 cm and a depth of 20–30 cm. Fresh and recently harvested seeds with a long dormant period are stratified in these trenches for autumn seeding or for further stratification in warm or cold trenches in boxes. After completely filling the summer trenches with a mixture of seeds, sand, or peat, a layer of thatch ranging from 10 to 15 centimetres is applied. Once every ten days, the mixture is stirred with a shovel and moistened as necessary. The ditches are encircled by a 50 cm-deep by 50 cm-wide troughs filled with water to prevent rat damage.

Stratification under the snow: It calls for the seeds to be packed into 1/3 full, light-textured bags. Many times, seeds are pre-moistened. The setup consists of spreading out the seed sacks flat, covering them with compacted snow, and finishing them with a layer of thatch or sawdust. Just before sowing, seeds are removed from beneath the snow and, if wet, drained.

For pre-treatment purposes, seeds can be divided into five groups as follows:

Group 1: Seeds requiring hot water treatment.

This group includes most leguminous trees that have pods and roughly flat, hard-coated seeds. Hot water treatment usually encourages these seeds to germinate more quickly and effectively. Here are the steps to follow. Once the water is nearly boiling, remove part of it from the fire. Spoon the seeds into a different container, pour in some boiling water, and soak for around 24 hours. Do not boil the seed. The seed will grow larger and eventually sink to the bottom of the pot as it collects water. Any swelled seeds should be taken out and immediately sown, and the remaining seeds should be kept in the pot for a further 24 hours. Once they have puffed up, they should also be planted (Fig.1).

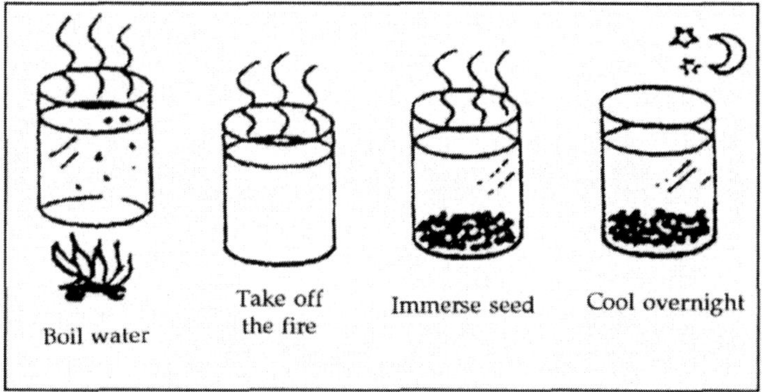

Fig. 1: Hot-water treatment of seeds

A fast and efficient way to pre-treat hard-coated seeds is to make small cuts in the seed coat with a knife or nail clipper. A small imprint is all that is required. Since nicking requires labor, it is advised only for tiny amounts of seed. Species such as Acacia spp., Cassia spp., *Leucaena leucocephala*, *Cassia spectabilis*, and *Parkinsonia aculeata* are included in this category. It is possible that these seeds will be stored for some time, and when preserving them, hot-water treatment is usually more important than when the seeds are fresh.

Group 2: Small light seeds with no need for pre-treatment

Many important species generate little, dry seeds, some of which can be stored for several months without losing their viability. Examples include *Juniperus procera*, Pinus species, *Iacaranda mimosifolia*, *Cupressus lusitanica*, Eucalyptus species, and Casuarina species. Additionally, certain species have little, dry seeds that don't require special handling yet spoil quickly. For these species, fresh seed germination yields the greatest results. Two examples are *Grevillea robusta* and *Markhamia lutea*.

Group 3: Large seeds with wings which should be removed before sowing

Many large-winged species are self-sufficient, but if the wings are cut off, the seed becomes easier to handle and takes up water faster after planting. It is best to disperse the fleshy seed as soon as the combretum fruit is opened. Some Terminalia species, such as Tipuana tipu seeds, can be stored for a while, whereas Combretum seeds must be sown immediately. Some examples of species are Terminalia spp., Combretum spp., and Tipuana tipu.

Group 4: Medium-sized or large seeds with no need for pre-treatment

This category includes several seeds with a high oil content, including Croton and Vitex species. Some of these seeds have a fruit pulp coating around them; it is best to wash the seed before drying or planting it. Some species, such as *Prunus africanus*, *Vitex* spp., *Syzygium* spp., *Bauhinia* spp., *Bridelia micrantha*, *Croton* spp., *Dovyalis caffra*, *Eriobotrya japonica*, *Erythrina abyssinica*, and Vitex spp., should be seeded as soon as possible for best results. On the other hand, seeds of Cordia species, *Melia azedarach*, *Schinus molle*, Sesbania species, and *Afzelia quanzensis* can be kept in storage for a long time. Albizia seeds can also be stored for a time, but only if they are kept free of insects. To separate the viable seeds from the non-viable ones that float, *Calodendrum capense* seeds should be floated in water to sowing. Some examples of species are *Afzelia quanzensis,* Albizia species, *Azadirachta indica, Bauhinia purpurea*, *Calodendrum capense*, Cordia species, Croton species, *Prunus africanus*, *Schinus molle*, *Sesbania* spp., *Syzygium* spp., and *Vitex* spp.

Group 5: Large seeds with hard seed coats which require cracking

It may be possible to mechanically crack the extremely hard seed coat of some seeds to allow water to enter the seed. However, extreme caution must be taken when cracking the seed coat so as not to harm the seed. Normally, this kind of seed can be kept in storage. *Adansonia digitata*, Podocarpus spp., and Ziziphus spp. are a few examples of species.

References

A Selection of Useful Trees and Shrubs for Kenya and Multipurpose Tree and Shrub Seed Directory, both published by ICRAF.

Becker, Robert & Grosjean, O. K. 1980 A compositional study of pods of two varieties of mesquite (Prosopis glandulosa, P. velutina). Journal of Agricultural and Food Chemistry, 28(1): 22–25.

Brown, F. J. & Belcher, Earl. 1979 Improved techniques for processing Prosopis seed.

Burkart, A. 1976a A monograph on the genus Prosopis (Mimosoideae). Journal of the Arnold Arboretum, 57(3): 219–249.

Burkart, A. 1976b A monograph of the genus Prosopis (Mimosoideae). Journal of the Arnold Arboretum, 57(4): 450–455.

Deno, N.C. 1993, Seed Germination Theory and Practice, 2nd Edn., US Department of Agriculture, USA.

FAO 1977 Report of the fourth session of the FAO panel of experts on forest gene resources (Canberra, Australia, March 1977), Rome, FO: FGR/4/Rep.

FAO 1980 Genetic resources of tree species in arid and semi-arid areas. FAO/IBPGR, 118 pp.

Ffolliott, Peter F. & Thames, John L. 1983 Taxonomy of Prosopis in Latin America. FAO/ IBPGR Project on Genetic Resources of Arid and Semi-Arid Zone Arboreal Species. FAO, Rome.

Flynt, T. O. & Morton, H. L. 1969 A device for threshing mesquite seed. Weed Science, 17(3): 302–303.

ISTA. 1976 International rules for seed testing. Seed Science and Technology, 4: 49–177.

Jonston, C. D. 1983 Ecology, control and identification of seed-infesting insects of new world Prosopis (Leguminosae). FAO/IBPGR Project on Genetic Resources of Arid and Semi-Arid Zone Arboreal Species. FAO, Rome.

Kingsolver, J. M., Johnson, C. D., Swier, S. R. & Teran, A. 1977 Prosopis fruits as a resource for invertebrates. In: Sympson, B. B. (ed.) Mesquite--its biology in two desert scrub ecosystems. Dowden, Hutchinson & Ross, Inc., Stroudsburg, Pennsylvania, pp. 108–122.

Moonshine Designs Nursery 2013, Seed Stratification Directions. [Online]. Available from: http://www.djroger.com/stratification.html.

Nelson, S. O., Bovey, R. W., and Stetson, L. E. 1978 Germination response of some woody plant seeds to electrical treatment. Weed Science, 26(3): 286–291.

Palmberg, C. 1981 A vital fuelwood gene pool is in danger. Unasylva, 33(133): 22–30.

Pfaffenberger, G. S. & Johnson, C. D. 1976 Biosystematics of the first-stage larvae of some North American Bruchidae (Coleoptera). USDA Technical Bulletin 1525, 75 pp.

Sharma, V.K., 2005. Propagation of Plants, Indus Publishing Company, New Delhi, India

Solbrig, Otto T. & Cantino, Philip D. 1975 Reproductive adaptations in Prosopis (Leguminosae, Mimosoideae). Journal of the Arnold Arboretum, 56(2):185–210.

Weber, F.R., Stoney, C. 1986. Reforesation in Arid Lands, Volunteers in Technical Assistance (VITA), Virginia, USA.

12

Pre-Storage Seed Treatments for Seed Invogoration

Priyanka Sharma

Concept: A few of the numerous environmental elements that affect seed quality maintenance are moisture, temperature, humidity, and storage conditions. When some seedborne diseases or pests like insects are taken into consideration, seed quality can still be negatively impacted. Applying one or more pesticides to seed is the most economical and efficient way to keep pests away and improve seed quality.

Because pesticides are harmful, handling treated seeds after application requires additional caution and safety measures.

Definition of Treated Seed

The definition of "treated" is "application of a pesticide or subject seed to a process intended to reduce, control, or repel insects, disease organisms, or other pests which attack the seed or seedlings."

Types of Seed Treatment

A. Pre sowing Seed Treatments

The treatments applied to the seeds prior to sowing enhance germination, increase their potential for vigor and keep the seeds healthy.

Pre sowing seed treatments includes the following.

i. Chemical procedures to enhance vigor potential and germination.

ii. Fungicidal and insecticidal agents.

iii. Special treatments

I. Chemical treatments to improve germination and vigour potential

Soaking / treating the seeds with nutrients vitamins and micronutrients etc.

Paddy: A 1% KCl solution can be steeped in seeds for a duration of 12 hours, which can enhance their germination and vigor potential.

Sorghum: To increase the potential for germination and vigor, seeds can be soaked in either 1% $NaCl_2$ or 1% KH_2PO_4 for a duration of 12 hours.

Pulses: To increase the germination and vigor potential of seeds, immerse them in a 100ppm solution of zinc, sulfur and manganese for four hours.

II. Insecticidal and Fungicidal Treatments

Seed health: It's a crucial component of premium seed. If a seed lot is contaminated by seed-borne illnesses and insect pests, farmers may find little use for it even if it meets tight requirements for germination, vigor, and purity. This is because such a situation could result in severe yield loss or possibly crop loss throughout a region.

Benefits of the insecticidal and fungicidal treatments

1. Prevents plant diseases from spreading.
2. Seed blights on seedlings and seed rot are prevented by it.
3. It promotes better seed germination.
4. It protects against insects that are stored.

5. It manages the soil insects

Seed Treatment Fungicides

Seed can be affected by bacteria, fungi, viruses, nematodes, and insects. It's critical to distinguish between pests and diseases that directly impact seeds, such as cereal bunt and *Tilletia* spp., and those that transmit through seeds, such as grain weevils, Tricoderma spp., and storage infections like *Aspergillus flavus*.

Before sowing, fungicides are sprayed on seeds to effectively defend against a variety of soil and seed-borne plant infections. Numerous seed rot and seedling blights that can occur during storage or after planting are prevented by chemical (fungicide) treatment. It is not a "cure-all" and won't keep diseases at bay during the growing season until the plants are self-sufficient. The prevention of loose smut through seed disinfection would be an exception to this rule.

Depending on the type of treatment and its intended use, fungicidal seed treatments can be categorized into three groups. (1) seed protection, (2) seed disinfection, and (3) seed disinfestation. There are multiple categories in which a particular fungicide can be used.

Seed disinfection - The process of disinfecting a seed involves eliminating a disease that has penetrated its living cells, been established, and spread—for example, loose smut of wheat and barley.

Seed disinfestations - Disinfestation is the process of reducing spores and other harmful organisms on the surface of the seed.

Seed protection - Chemical treatment is used in seed protection to keep pathogenic soil organisms away from seeds and young seedlings.

One of four forms— The common methods for applying seed treatment chemicals are dust, slurry (a solution of wettable powder in water), liquids, or planter-box formulations.

Based on their chemical makeup, seed treatment fungicides can be categorized as organic or inorganic, metallic, or non-metallic, and, until recently, mercurial, or non-mercurial. Fungicides used to treat seeds were typically divided into two categories: volatile and non-volatile, prior to the cancellation of the "volatile mercurials" Following the removal of the volatile mercurials, most fungicides that are currently authorized for use on seed are categorized as non-volatile. To achieve effective control while employing this kind of material, the seed must be completely covered. It is now possible to utilize some of the relatively recent systemic insecticides as seed treatment agents. It has long been known that compounds that may migrate inside seeds or plants to suppress pests are desirable. We refer to these materials as "systemic." When applied as directed by the manufacturer, it acts systemically across the host plant, suppressing or delaying the growth of fungi and insects without changing the metabolic processes of the host.

Seed Treatment Insecticides

Insecticides are often sprayed on seed to control or reduce insect damage to seed during storage, as well as to avoid damage from insects such as wireworms and seed corn maggots in the soil.

Combinations

Since many pesticides are selective, it's usual practice to combine two or more compounds in the treater tank or use a separate tank to offer the required range of control.

Seed processors can now choose from combinations offered by pesticide producers; however, before blending two or more pesticides, the compatibility of the ingredients needs to be established, since certain combinations might significantly lower seed germination. Additionally, the order in which we administer two or more pesticides—even at different times—may be crucial. If the seed is to be treated with a single pesticide or a combination, carefully read the label and adhere to the manufacturer's instructions.

Formulation of Fungicides /Insecticides

Fungicides / insecticides are available in the form of dusts, wettable powders and liquids.

1. **Dusts:** 200–250 g/q of seeds are typically applied. The primary drawback is the presence of dust both during and following seed treatment.
2. **Slurry:** A soap-like water suspension is added when administering this type of fungicide to seeds, and the seeds can be mixed with the suspension using a specialist slurry treater.
3. **Liquids:** Here a fungicide treatment is applied to the seeds, and they are thoroughly mixed with them. E.g.: Panogen and mercuran.

Safety

Chemicals that are safe for users and the environment are typically used. Highly persistent fungicides like hexachlorobenzene (HCB) and extremely poisonous substances like organic mercurials (Ceresan and others) are being replaced by new chemicals, these substances have historically resulted in serious poisoning incidents, some of which were fatal. Most of these incidents happened because of treated seed being fed to cattle or consumed by humans rather than being planted. The following safety measures still need to be followed, even with the new, less hazardous substances.

- Seed treatment ought to be done in a well-ventilated area.
- Treated seed needs to be properly labelled and should never be used for food or feed. It is necessary to prevent coming into touch with chemicals by skin contact and inhaling dust. Wearing protective gear is recommended.
- Empty containers, like empty pesticide containers, should never be repurposed in a home or on a farm. They should be disposed of appropriately.

III. Special treatments

i) Seed hardening treatment

Seeds can be hardened for 2 purposes

I) Drought tolerance

II) Cold tolerance

The major purpose of the treatments is to help the seeds withstand the initial cold and drought. Only seeds from temperate crops and trees are treated with cold tolerance once they have germinated.

The most important factors to be considered while seed hardening are

- Seed: to-solution (1:1) ratio
- The duration of the soaking
- The drying process.

The way the seed hardening procedure is carried out determines how effective the therapy will be. The volume of solution should never exceed the quantity of seeds. The seedlings should absorb every added solution. Any remaining solution should be removed because it has a leaching impact. The process of germination begins when the seeds encounter water. The seeds ought to be dried to their initial moisture content at the conclusion of the soaking time. When these seeds are sown, germination will finish sooner than it does for non-hardened seeds, which take longer to germinate. E.g.: $CaCl_2$, KCl, KH_2PO_4.

ii) Seed fortification

The primary goal is to nourish seeds. Achieving strong vigour to overcome adverse soil reactions is the main objective. For example, seed fortification with 0.5–1% MnSO4 will increase the seeds' capacity for oxidation and reduction, which will eventually result in increased germination.

iii) Moist sand conditioning

The dosage of this need-based treatment can be increased by 2–4%. The solution to water ratio should be 1:1, however you can use a little more than that. Soybean and other proteinaceous seeds shouldn't be soaked in water. For these seeds, combine the seeds with 5 to 10% MC of damp sand. It must be retained for the allotted amount of time. The process is called hydration of damp sand.

iv) Seed pelleting

Here, the seeds have a layer of nutrition on them. In forest tree seeds, this method is often used.

Importance

- Generally, this method is used for small sized seeds.
- Pelletizing allows for the enlargement of the seed's size and creation of a free-flowing seed.
- This will allow us to lower the seed rate.
- In tree seeds, it is also crucial for aerial sowing (gum arabica).

Materials used: Nutrients, adhesive, filler material.

Inert materials: Lime, $CaCO_3$, Chalk powder.

Plant products: It has been discovered that neem, notchi, and arappu (*Albizia amara*) are good sources of saponin, a growth stimulant that functions similarly to GA.

v) Seed Infusion

Nutrition and growth-promoting material infusion involves using organic solvents such as dichloromethane and acetone. The compounds in the seed are gradually increased by the organic solvents. Using this approach, the seed materials can be dried without losing their original moisture content. Because the organic chemicals are evaporative, all we need to do is let the seeds sit in a dry place for five to ten minutes after the infusion is finished. During this time, the organic solvents will evaporate, allowing us to proceed with the sowing process. Another method for thawing seed dormancy is seed infusion.

vi) Osmotic Priming

This process is essential even though it is very expensive, especially for large-seeded legumes like beans, peas, etc. They are prone to soaking harm, have a big embryo, and a high protein content. Seeds with high protein content are hydrophilic and hygroscopic. Osmotic priming only requires gradually letting the seeds absorb water. Polyethylene glycol (PEG) solutions are used as osmotic solutions. Manitol is extremely poisonous. PEG will gradually raise the water content of seeds because it is inert. Preconditioning through osmotic priming energizes the seeds, resulting in uniform, early, and higher field emergence along with better vigor in the seedlings.

vii) Fluid Drilling

This technology was developed specifically for mechanical sowing of seeds, especially ones that have germinated. Guar gel, a type of jelly, is applied on the seeds. It serves as a buffer to prevent germination-related harm to the seeds during sowing.

viii) Separation of Viable Seeds

It's a crucial concept, especially for groundnuts. This is a useful technique to get the required plant population and seed germination. The real population requirement for groundnuts is thirty plants per square metre. In ground nuts, the actual seed multiplication rate is 1:8. We will be able to sustain the necessary plant population in the field once the approximately 30–40% of dead seeds are removed. This is the foundation for this technology's evolution.

This can be done in 2 ways.

1. Manual separation (groundnut) based on radicle emergence.
2. The process of Incubation, Drying, and Separation (IDS)

B. Pre Storage Treatments

The main goal of pre-storage treatments for harvest-fresh seed is to prevent senescence from degrading while the seed is being stored. Prescription pre-storage seed treatments can help protect seed storage, which is once more at risk from disease and insect attack.

i. Halogenation
ii. Antioxidant treatment
iii. Seed sanitation

C. Mid Storage Treatments

During senescence, seeds stored in storage incur damage to their cell membranes. In addition to improving the productivity and seed viability of stored seeds, mid-storage seed treatments can partially reverse the effects of ageing on seeds and restore some of their vigour.

i) Hydration – Dehydration

Low- and medium-vigor seeds are soaked in water for short periods of time—either with or without additional chemicals—to raise the moisture content to 25 to 30%. The seeds are then dried to a safe level before being stored in a dry environment.

The hydration – dehydration treatments

1. Only applied to seeds that have been stored.
2. The moisture equilibration and moist sand conditioning treatments, which allow moisture to be absorbed by the seed gradually, are advised for relatively high-vigor seeds, pulse seeds, and seeds of leguminous vegetable crops.
3. Is effective for low and medium vigor non-leguminous seeds.
4. Leguminous seeds shouldn't be soaked directly.
5. Would not revive a seed that has already lost its viability.

Types of H-D treatments

Among the wet treatments are soaking, dipping, spraying, stepwise hydration, soaking and conditioning, drying with moisture equilibration, and drying with moisture. The traits of the seed and its first state of vigour determine which treatment is best.

Soaking – Drying (S-D)

After being covered with water or a chemical solution, stored seed is allowed to sit at room temperature for two to six hours, depending on the variety, and is periodically stirred.

After the soaked seed is removed, it is allowed to dry on the surface in the shade for a while until it regains its original moisture content. Chemicals that can be employed at concentrations of 10^{-4} to 10^{-3} M include sodium or potassium phosphate (di and mono basic), sodium chloride, p-hydroxy benzoic acid, p-amino benzoic acid, oxalic acid, potassium iodide, etc. in diluted solutions. It is also possible to add formulations that are fungicidal and insecticidal to the soak water.

Dipping – Drying (D-D)

To absorb surface water and then dry back in S-D, seeds are dipped in water or solutions of the aforementioned chemicals for a short period of time—two to five minutes. The wet seed is then removed immediately and covered for two to six hours, depending on the material. Most of the summer and winter vegetable seeds with high and high-medium vigor, such as wheat, rice, and jute, react favourably to this treatment.

Spraying – Drying

After spreading out a thin layer of seeds, water (about 1/5 to ¼ of the seed weight) is sprayed on it in two equal portions, flipping the seed layer after the first spray. The seed layer is then covered with a polythene sheet and allowed to dry for a couple of hours before being exposed to more water. This treatment works well and is appropriate in a way that is comparable to D-D.

Moisture Equilibration – Drying (ME – D)

Within a closed, moisture-saturated chamber with interior wet blotters that offer an estimated humidity of 100% at room temperature, the seeds are distributed in thin layers on trays that are kept on an elevated platform. Depending on the substance and outside temperature, the seed is dried again in a 24-48-hour period. This therapy, which causes a gradual rise in moisture content, works wonders for soaking-injured seeds. However, ME-D is challenging to apply widely, and it is not recommended for low-vigor non-leguminous seeds due to the treatment's potential for ageing, particularly when used for extended periods of time.

Moist Sand Conditioning – Drying (MSC-D)

Though less complicated to use, this treatment is comparable to the moisture equilibration procedure. Three times as much air-dry sand is used as seed in order to completely mix the seed with pre-moistened sand for gradual and progressive moisture uptake. Sand that has been previously cleaned and dried to fine grain building grade is given the necessary amount of water or chemical solution to bring the moisture content of the sand down to 5–10. Water additions should be adjusted such that they provide the necessary hydration effect without starting the germination process. Depending on the material and sand moisture content, the mixture is left to sit at room temperature for 16 to 36 hours after the dry seed and premoistened sand are combined. Sand is absorbed by the seed, which then becomes hydrated. After an incubation period, the hydrated seed is sieved off the sand and dried to its initial weight.

Mode of Action

The major objective of hydration is to increase the moisture content of seeds to 25–30% (wet weight basis) so that they can be dried to safe levels for storage in the dry. Through the regulation of free radical reactions and the ensuing peroxidative damage to lipoprotein cell membranes, the hydration-dehydration treatment may enhance vigor.

Seed Treating Equipment

Commercial seed treaters are produced to precisely measure and administer the right amount of insecticide to a specific weight of seed. In general, there are three different kinds of commercial seed treaters available on the market: direct treaters, such as Mist-O-Matic and Panogen, and slurry treaters.

Dust Treater *(Gustafson XL Dry Powder Seed Treater)*

Controlling the Flow of Seed

The weigh pan sits right beneath the hopper on top of the treater. The number of seeds can be adjusted that flows into it by turning the hand wheel on the side of the hopper. The hopper's scale indicates the amount of open space (in inches) between the gates. Gates should be left open for however many inches it takes to hold the necessary numbers of pounds per dump in the weigh pans when they tilt in either direction. By properly adjusting the counterweight up or down on the counterweight arm, the number of pounds per dump can be changed.

Powder Application

To make sure that the appropriate amount of powder is being applied to the seed flow, a preliminary test that involves running through a specified number of pounds of seed (for example, 100 pounds) via the feeder is required. The measuring cup that comes with the feeder should be used to collect the powder as it comes off the vibrator throughout this run. Weighing the powder can you establish how much is being put to that amounts of seeds once the specified amount has been seeded. After that, the vibrator speed can be modified appropriately. After that, one or more additional tests should be conducted to find the ideal vibrator speed setting for accurate coverage.

Approximate Setting

No. Dumps	Powder Scale Opening	Syntron Setting	Oz. Produced/100 lbs.
25	1/2	60	2
25	3/4	60	5
25	3/4	70	6
25	3/4	80	7
25	1	60	10

The counterweight arm's number four delivers five pounds per dump.

2. Slurry Seed Treater

The slurry treatment principle states that wettable powder treatment material is suspended in water. The treatment material provided as a slurry is precisely measured using a simple equipment constructed of a seed dump pan and slurry cup. The cup adds a specific amount of slurry to a mixing chamber where the seeds are mixed with each dump of seeds.

Even though using the slurry treater is quite easy, it's important to fully comprehend all of the different operation processes.

1. The metering principle, which involves adding a certain quantity of slurry to a specific weight of seed, is the same for direct, ready-mix, and fully automatic treaters.
2. To reach a certain dump weight, slurry treaters are equipped with a seed gate that controls the flow of seed to the dump pan. A consistent dump weight for a given can be achieved with the appropriate seed gate configuration.
3. The size of the slurry cup or bucket and the slurry concentration determine how much treatment material is applied. There comes a point at which the dump pan overbalances the counterweight and spills into the mixing chamber as it fills. This initiates a mechanism to add a cup of slurry to the mixing chamber and positions the backup weighing pan to receive the seed inflow. For this reason, for every deposit of seed, one cup of slurry is added.
4. An auger-style agitator within the mixing chamber combines and transfers seed to the bagging end of the chamber. The auger's speed matters because a more uniform distribution is achieved at slower rates.
5. Slurry tanks can hold anything from 15 to 35 gallons, depending on how big the treater is. They have agitators installed, which combine the slurry in the tank and maintain its suspension while the machine is working. Before treatment, it is crucial that the powder be completely suspended in water. It is necessary to clear out the sediment from the slurry cups' bottoms if the treater has been inactive for any length of time.
6. Slurry cups of the appropriate size must be used. Nowadays, most machines include cups with rubber plugs and ports for 15 cc, 23 cc, and 46 cc capacities. Certain users opt to combine the slurry in a separate tank and subsequently move it to the slurry chamber when required.

1. Direct Treaters

The Panogen and Mist-O-Matic treaters are examples of more modern direct treaters. Originally, these two were intended to apply liquid treatment that was not diluted. As opposed to slurry treaters, which apply 23 cc of material per 10 pounds of wheat, they apply 14 to 21 cc (1/2 to 3/4 ounces) each bushel of "wheat." Only slightly volatile liquid compounds that don't need full, uniform covering to work well are appropriate for this tiny amount of substance. Subsequent adjustments for direct treaters include dual tanks that allow insecticide and fungicide to be added simultaneously, as well as improvements for slurry treatment. Since metering is achieved by timing a treatment cup and seed dump, the metering equipment utilized in both types of direct treaters is comparable to that of the slurry treater. Other than that, there are clear differences between the two direct treaters and the slurry treater. To change the weight of the seed dump, these two direct treaters each include an adjustable dump pan counterweight. When using slurry treaters, this is not feasible.

2. Panogen Seed Treater

The Panogen treater functions in a comparatively straightforward manner. One cup of treatment is metered with each seed pan dump by a tiny treatment cup that operates from a rocker arm that is immediately off the pan and out of a little reservoir. A tube carries fungicide to the head of the rotating drum seed mixing chamber. As the seed passes through the rotating drum, it rubs against the seed as it pours in with the seed from the dumping pan. By varying the weight of the seed dump and the size of the treatment cup, the optimum treatment rate can be achieved. Treatment cup sizes are not based on real size; for example, a 3/4-ounce cup applies 3/4 ounce (22.5 cc) of treatment per bushel with six dumps per bushel. Treatment cup sizes are determined by the treating rate in ounces. This cup's real dimensions are roughly 3.75cc.

3. Mist-O-Matic Seed Treater

The "mist-o-matic" treater mists the seed directly with the therapy. The "Panogen" treater's metering mechanism is comparable to that of the treatment cups and seed dump. Cup sizes are denoted by the exact count of counts they hold, such as 2.5, 5, 10, 20, and 40. To maintain the level in the tiny reservoir from which the treatment cups are fed, the treater is outfitted with a big treatment tank, a pump, and a return. Following metering, the treatment material is fed onto a fluted disc that rotates quickly and is positioned beneath a cone that disperses seeds. To coat seed that falls over the cone through the treating chamber, the disc break up droplets of the treatment solution into a fine mist and sprays it forth. Two movable retarders placed just beneath the seed dump are intended to provide a constant seed flow across the cone in between seed dumps. This is significant because material from the rotating disc is constantly dripping. By choosing the appropriate treatment cup size and adjusting the seed dump weight appropriately, the required treating rate can be achieved.

13

Classification of Stored Pests and Insects and Its Role in Cereals, Pulses and Oilseeds Crops

Abhilisa Mudoi

Stored grain pests are one of the major constraints in obtaining quality seeds. They cause substantial yield loss in storage conditions affecting both the quality and quantity of the produce. Apart from causing direct loss to the grains, they also make them unfit for consumption through their excreta, scales and wings resulting in development of unpleasant odour (Upadhyay and Ahmad, 2011). They also act as a source of medium for growth of fungi or bacteria (Daglish *et al.*, 2018). The method of infestation of stored gain pests is different from the insect pests infesting crops under field condition. Generally, the infestation of storage pests starts after harvesting. It has been reported that in India, there is a loss of about 14 million tons worth Rs. 7000 crores due to the infestation by different storage pests (Chaubey, 2022). The post-harvest loss in terms of quantity and quality by different storage insect-pests accounts for about 10-30 per cent (Rajasri and Kavitha, 2015; Yaseen *et al.*, 2019).

Cereals, pulses, and oilseeds are major food source to a wide range of population in India. Out of them, cereals like rice, wheat, maize, oats, barley etc. is the staple food for majority of Indians. Cereals, pulses, and oilseeds provide protein, fat, fibre and micronutrients such as vitamins, iron, zinc, phosphorus etc (Oso and Ashafa, 2021). As a result of increasing population, developing country like India has faced a major challenge in obtaining safe and quality foods. Storage pests pose a major threat in obtaining quality cereals, pulses and oilseed crops which are the major grown crops in the country (Bhargava *et al.*, 2007). In pulses, it has been reported an annual post-harvest loss of around 25–50% mostly due to infestation of insect-pests under storage condition (Yaseen *et al.*, 2019). The favourable climatic condition of India is suitable for growth and development of these storage pests. Their infestation results in loss of germination and nutritive value of seeds (Yaseen *et al.*, 2019).

Classification of Stored Grain Insect-Pests

1. Based on insect order, stored grain pests could be placed under:
   a. Coleopterans: E.g. Pulse beetle, red flour beetle, lesser grain borer, rice weevil etc.
   b. Lepidopterans: E.g. Angoumois grain moth, rice moth, almond moth, Indian meal moth etc.
2. Based on feeding behaviour, they are classified as
   a. External feeders: These insect pests feed from the surface of grain. E.g. Red flour beetle, khapra beetle, almond moth, rice moth, Indian meal moth
   b. Internal feeders: These insect pests feed within the kernel of seed. E.g. Rice weevil, pulse beetle, lesser grain borer, cigarette beetle, potato tuber moth
3. Based on type of infestation, they are grouped as
   a. Primary pests: Attack whole and undamaged grains. E.g. Pulse beetles, weevils.
   b. Secondary pests: Attack broken or already damaged grains. E.g. Red flour beetles

Important insect-pests of stored produce:

Sl. No	Common name	Scientific name	Order	Family
1	Angoumois grain moth	*Sitotroga cerealella*	Lepidoptera	Gelechiidae
2	Rice moth	*Corcyra cephalonica*	Lepidoptera	Pyralidae
3	Almond moth	*Ephestia cautella*	Lepidoptera	Pyralidae
4	Indian Meal Moth	*Plodia interpunctella*	Lepidoptera	Pyralidae
5	Rice weevil	*Sitophilus oryzae*	Coleoptera	Curculionidae
6	Lesser grain borer	*Rhyzopertha dominica*	Coleoptera	Bostrichidae
7	Red Flour Beetle	*Tribolium castaneum*	Coleoptera	Tenebrionidae
8	Pulse beetle	*Callosobruchus chinensis*	Coleoptera	Bruchidae
9	Khapra beetle	*Trogoderma granarium*	Coleoptera	Dermestidae
10	Cigarette or tobacco beetle	*Lasioderma serricorne*	Coleoptera	Anobiidae
11	Drug store beetle	*Stegobium paniceum*	Coleoptera	Anobiidae
12	Saw toothed beetle	*Oryzaephilus surinamensis*	Coleoptera	Silvanidae
13	Sweet potato weevil	*Cylas formicarius*	Coleoptera	Curculionidae

(Upadhyay and Ahmad, 2011; Parimala *et al.*, 2013; Yaseen *et al.*, 2019)

1. Angoumois Grain Moth

Host: Wheat, Barley, Jowar, Maize, Paddy

Nature of damage

- Larvae are destructive and feed on grain kernels.
- Larva bores into grain and feeds inside up to 30 – 50 percent seed is damaged.
- Sometimes whole grain is damaged.
- Infestation confined to upper 30 cm depth.
- Damaged grain emits unpleasant smell and gets covered with scales.

2. Rice Moth

Host: Sorghum, Groundnut, Maize, Paddy

Nature of damage

- Caterpillar alone is responsible for damage. It prefers partially damaged grains and feed.
- It pollutes food grains with frass, moults, and dense webbing. In case of whole grains, kernels are bound into lumps.
- Grains are converted to webbed mass.
- Damaged grain / flour with bad odour unfits for consumption.

3. Almond Moth

Host: Cereals, Dry fruits like almonds, walnut, dates, Dry garlic bulb

Nature of damage

- The caterpillars make tunnel into the food material.
- The number of silken tubes is sometimes extremely high and these clog the mill machinery where the infested grains have been sent for milling.

4. Indian Meal Moth

Host: Soybean, Dry fruits, Nuts, herbs, dead insects

Nature of damage

- Larva causes serious damage and contaminates the grain with excreta, cast skins, webbings, dead individuals, and cocoons; prefers to eat the germ portion and hence grains lose viability.

- It feeds superficially but may construct more than one silken tunnel.

5. Rice Weevil

Host: Jowar, Wheat, Maize, Paddy

Nature of damage

- Grubs bore into grain and feed internally and destroy more than they eat.
- Hollowed out grains.
- Kernels reduced to powder.
- Grain Heating

6. Lesser Grain Borer

Host: wheat, Sorghum, Maize, Paddy

Nature of damage

- Both grubs and adults are destructive, feeding inside the grains and adults come out leaving irregular hole. They spoil more than they eat.
- 1st instar larva bore inside the grains and later instars gets carved and depend on flour produced by adults.
- Kernels reduced to shells and milled products into mass of excreta.
- Grain heating

7. Red Flour Beetle

Host: wheat flour, Cereal products, Pulses, Dry fruits

Nature of damage

- Both grubs and adults feed on already damaged grains and can't feed on whole grains.
- Damage is serious during monsoon.
- Flour turns greyish and mouldy, has pungent odour.

8. Pulse beetle

Host: Cowpeas, Peas, Lentil, Green gram

Nature of damage

- The grubs could feed the entire seed content leaving only the shell.
- Damaged grain is unfit for consumption.
- Damaged grain converted to flour by traders give off flavour.

9. Khapra Beetle

Host: Wheat, Rice, Gram, Maize

Nature of damage

- Grub and adult feed the grain starting with germ portion, surface scratching and devouring the grain.
- Damage is confined to upper (50 cm) layers of heap and are resistant to temperature, humidity, and starvation.
- Infestation is indicated by presence of cast skins, frass and hair on bags.
- It reduces grain into frass. Excessive moulting creates public discrimination, loss of market appeal due to presence of cast skins, frass, and hair. Crowding of larvae leads to unhygienic conditions in warehouses.

10. Cigarette Beetle

Host: Tobacco products, Dried black pepper, Dried red pepper, peas

Nature of damage

- Both grubs and adults bore holes into tobacco products like cigarettes, cheroots (cigars) and chewing tobacco.
- It also damages stuffing furniture.
- Grubs bore through cardboard boxes and other packaging material in search of place to pupate.

11. Drug Store Beetle

Host: Flours, Breads, cookies, chocolates, Non-food material like wool, leather

Nature of damage

- It tunnels into stored products like turmeric, ginger, coriander and dry vegetable and animal matter and make them unfit for consumption.

12. Saw Toothed Beetle

Host: Dry fruits, nuts, candy, Dried meat, Tobacco

Nature of damage

- Both grubs and adults cause roughening of grain surface producing off odour
- Grains with higher percentage of brokens and foreign matter attract heavy infestation, which leads to heating of grain.

13. Sweet Potato Weevil

Host: Sweet potato

Nature of damage

- Thickening and malformation of vines and often cracking of the tissue.
- Discoloration, cracking, or wilting of damaged wines.
- An infested tuber is often riddled with cavities or tunnels.
- Attacked tubers become spongy, brownish to blackish in appearance.
- Start rotting form the top and develop an unpleasant smell and a bitter taste.
- Unfit for human consumption.

Stored Grain Pests Infesting Cereals

Stored grain insect pests cause extensive damage to cereals. About 20-30 per cent loss has been reported as a result of infestation by different insect-pests in cereals (Chomchalow, 2003). The cereals are infested by both primary and secondary pests (Parimala *et al*., 2013). In some cases, the infestation of stored grain insect pests starts from the field itself. The major stored grain insect-pests infesting cereals are Rice weevil (*Sitophilus oryzae*), Grenary weevil (*Sitophilus granarius*), Angoumois grain moth (*Sitotroga cerealella*), Indian meal moth (*Plodia interpunctella*), Rice moth (*Corcyra cephalonica*), Lesser grain borer (*Rhyzopertha dominica*) (Yaseen *et al*., 2019). The nature of damage of these pests are discussed earlier.

Stored Grain Pests Infesting Pulses

Pulses are a good source of protein along with fibre providing carbohydrates, starch, vitamins, minerals etc. Pulses provide various essential amino acids (methionine, cysteine, threonine, tryptophan, and lysine) to human health (Hall *et al.*, 2017; Venkidasamy *et al.*, 2019). Pulses are the main source of protein in developing countries as compared to animal protein. It has been reported that the extent of post-harvest loss in pulses range from 25-50 percent (Ngamo *et al.*, 2007; Yaseen *et al.*, 2019). The important stored grain pest of pulse are the bruchids, *Callosobruchus* spp. which results in about 5-10 per cent loss in storage (Ngamo *et al.*, 2007). Gbaye *et al.*, 2011 reported that the bruchids may even cause 100 per cent damage to pulses under storage. The bore holes created by bruchid results in loss of germinative ability of the seeds as well as it makes them unfit for consumption. Apart from the bruchids, pulses are also infested by Cigarette beetle (*Lasioderma serricorne*), khapra beetle (*Trogoderma granarium*) etc (Yaseen *et al.*, 2019). The nature of damage of these pests have already been discussed above.

Stored Grain Pests Infesting Oilseeds

Oilseeds like mustard, rapeseed, groundnut, sunflower etc. are grown throughout the world as they are a good source of vegetable oils. About 20-40 per cent oil could be extracted from these crops. They are good sources of B-vitamins, vitamin A, thiamine, nicotinic acid etc. (Among the stored grain pest, the oilseeds are infested by Indian meal moth (*Plodia interpunctella*), cigarette beetle (*Lasioderma serricorne*), khapra beetle (*Trogoderma granarium*) etc. (Yaseen *et al.*, 2019). (Please refer to the nature of damage mentioned above).

Rodent Infestation in Storage

Rodents are one of the most harmful pests of stored grains. It causes both direct and indirect damage to stored grains. It makes the grains unfit for consumption through its droppings and body hairs. It affects both the quality and quantity of produce. The infestation by rodents is further accelerated by development of mould and bacterial growth (Kumawat *et al.*, 2016). The rodent infestation results in further spread of diseases in human beings. The infestation of rodents starts not only in storage condition but also during production, processing and transportation. As a result of their widespread occurrence and ability to reproduce in large numbers, they could cause tremendous loss to crop in field as well as in storage. Several rats and mice are available in northeastern states belonging to the genus *Rattus*, *Bandicota*, *Mus*, *Cannomys* etc (Kumawat *et al.*, 2016). In Assam, five species of rodents viz., *Rattus rattus*, *R. sikkimensis*, *R. norvegicus*, *Bandicota bengalensis bengalensis* and *Mus musculus castaneus* were reported from diffent storage godowns of Assam (Thangjam and Dutta, 2012).

References

Bhargava, M.C., Choudhary, R.K. and Jain, P.C., 2007. Advances in management of stored grain pests. Entomology: Novel Approaches. PC Jain and MC Bhargava (eds), pp.425-451.

Chaubey, M.K., 2022. Chapter-3 Biology of Common Stored-Grain Insect Pests. Latest trends in zoology and entomology sciences. p.31.

Chomchalow, N (2003). Protection of stored products with special reference to Thailand. AU J Tech 7(1):31–47.

Daglish, G. J., Nayak, M. K., Arthur, F. H. and Athanassiou, C. G. (2018). Insect pest management in stored grain. Recent advances in stored product protection. 45-63.

Gbaye, O.A., Millard, J.C. and Holloway, G.J. (2011) Legume type and temperature effects on the toxicity of insecticide to the genus Callosobruchus (Coleoptera: Bruchidae). J Stored Prod Res 47:8–12.

Hall, C., Hillen, C., & Garden Robinson, J. (2017). Composition, nutritional value, and health benefits of pulses. Cereal Chemistry. 94(1): 11-31.

Kumawat, M. M., Tripathi, R. S., Riba, T., Pandey, A. K. and Rao, A. M. (2016). Rodent Pests of Arunachal Pradesh and Their Management. Technical bulletin (34).

Makundi, R. S., Kilonzo, B. S. and Massawe, A. W. (2003). Interaction between rodent species in agro-forestry habitats in the western Usambara Mountains, north-eastern Tanzania, and its potential for plague transmission to humans.

Ngamo, T.S., Ngatanko, I., Ngassoum, M.B., Mapongmestsem, P.M. and Hance, T. (2007). Persistence of insecticidal activities of crude essential oils of three aromatic plants towards four major stored product insect pests. Afr J Agric Res. 2(4):173–177.

Oso, A. A. and Ashafa, A. O. (2021). Nutritional composition of grain and seed proteins. Grain and seed proteins functionality. 31-50.

Thangjam, B. and Dutta, B. C. (2012). Species composition of rodents in rural and commercial food grain storages in Assam. Indian Journal of Entomology.74(2): 108-113.

Venkidasamy, B., Selvaraj, D., Nile, A. S., Ramalingam, S., Kai, G. and Nile, S. H. (2019). Indian pulses: A review on nutritional, functional and biochemical properties with future perspectives. Trends in Food Science and Technology. 88: 228-242.

Yaseen, M., Kausar, T., Praween, B., Shah, S.J., Jan, Y., Shekhawat, S.S., Malik, M. and Azaz Ahmad Azad, Z.R. (2019). Insect pest infestation during storage of cereal grains, pulses and oilseeds. Health and Safety Aspects of Food Processing Technologies, pp.209-234.

Parimala, K., Subramanian, K., Mahalinga Kannan, S. and Vijayalakshmi, K. (2013). Seed storage techniques–a primer. CIKS, Chennai, India.

Upadhyay, R. K. and Ahmad, S. (2011). Management strategies for control of stored grain insect pests in farmer stores and public ware houses. World Journal of Agricultural Sciences. 7(5): 527-549.

14

Seed Storage Godown-Sanitation, Fumigation and Storage Insect Pest Control Through Physical, Chemical and Biological Methods

Abhilisa Mudoi

Sanitation

Sanitation and maintaining good hygiene conditions are the important aspect of seed storage in storage godowns. Maintenance of good hygiene conditions will maintain the quality of produce in the long run (Mahapatra *et al.*, 2014) and these practices should not be overlooked. The threshing floors should be cleaned properly before the operation so that there is no mixing of seeds or any insect-pests or pathogens. Harvesting and threshing should be followed by cleanliness of godowns to protect the seeds from the harmful pest and pathogens (Upadhyay and Ahmad, 2011). The storage godowns should be well lighted and the area should be preferably sunny, free from weeds and mostly situated at elevated areas. Damp areas should not be selected for building storage godowns as it will affect the moisture per cent of seeds thereby enabling easy growth of insect-pests and diseases (Parimala *et al.*, 2013). Before bringing the new produce to the storage godowns, they should be cleaned properly. The cracks and crevices of storage godowns if any should be sealed ahead and the walls of the godowns should be whitewashed (Hazam and Kumar, 2022). The new seed lots should not be mixed with the older lots. The floors should be disinfected, and it should be cleaned regularly.

Depending on the type of seeds, different storage facilities are available which varies from place to place. Mostly the grains are stacked properly after putting them in storage bags of definite sizes (Guru and Mridula, 2021). Before putting the grains in bags, the bags should be cleaned properly and to be ensured that they are free from any holes. The bags should be inspected from time to time to avoid infestation of insect-pests and diseases (Khan *et al.*, 2012) If it shows any sign of infestation, the bag should be immediately discarded to avoid further spread of insect-pests and diseases to other seed lots. The bags should not be kept directly at the floors of storage godowns as the insect-pests and rodents would directly infest them (Khan *et al.*, 2012). They should be always kept in racks vertically with the provision of proper aeration amongst them.

Fumigation

It is one of the popular and widely used technique in storage godowns against storage insect pests. Toxic gases are emitted in confined environmental conditions and insects are exposed to this gaseous environment (Zettler and Arthur, 2000; Parimala *et al.*, 2013). These harmful gases enter the insect body through spiracles to reach the trachea (Upadhyay and Ahmad, 2011). It causes disruption of tracheal system thereby bringing mortality of these pests (Kedia *et al.*, 2015). Fumigation should be undertaken with utmost care. While fumigating, the doors and windows of the storage godowns should be closed to avoid mixing of harmful gases to the outer environment (Kedia *et al.*, 2015). Along with these, the dead insect-pests should be removed from the storage godowns in the next day itself and they must be disposed safely. Several fumigants were used earlier to manage stored grain insect-pests. However, considering the health hazard, environmental pollution, and regulatory aspects, most of the earlier used fumigants are banned (Zettler and Arthur, 2000).

1. Physical Control Methods

a. **Cleaning of seeds:** After harvesting and threshing operations, seeds must be properly cleaned to ensure the seeds free from dust particles because the dust particles provide a means for maintaining the life cycle of storage pests. The threshing floor should be cleaned properly to avoid mixing of weed seeds or other infested seeds. Sieving and winnowing are one of the popular methods mostly followed by farmers to clean the fresh seeds. Before bringing the fresh seed lot to storage godowns, the old stock should be safely disposed to prevent their admixtures. The walls of storage godowns should be cleaned properly, preferably whitewashed and it should be free from cracks and crevices (Parimala *et al.*, 2013).

b. **Sun drying:** Sun drying of seeds is very important to maintain optimum moisture content of seeds. High seed moisture results in increase infestation of the produce from stored grain insect pests. Solar drying reduces the moisture content of seeds to safe storage levels to prevent them from storage insect-pests and thereby enhancing the storage life

(Asemu *et al.*, 2020). It has been reported that most of the infestation of stored grain pests could be avoided if the moisture content of the seeds is maintained within 8-12 per cent.

c. **Alteration of temperature of seeds:** Alteration of temperature is an effective way to arrest the growth and development of stored grain insect pests. It has been reported that increasing the temperature of pulses to 60-65°C or lowering it below 12°C could effectively kill the egg, larval and pupal stage of pulse beetle (Upadhyay and Ahmad, 2011).

d. **Use of inert dusts, sand, and silica:** Mixing of seeds with these materials brings mortality of stored grain insect-pests through desiccation and by lowering the relative humidity. Sand provides a protective coating on the surface of seeds. Now-a-days, diatomaceous earth and silica are also used against stored grain insect-pests (Banks and Field, 1995; Upadhyay and Ahmad, 2011).

e. **Radiation treatment:** Generally, gamma radiation is widely used for treating seeds which provides protection against stored grain insect pests. Radiation brings about sterilization of stored pests (Kells *et al.*, 2001). The seeds are given radiation treatment in closed chambers without impairing the nutritional content of seeds (Fawki *et al.*, 2014).

f. **Use of traps:** Trap technology is widely used against stored grain pests. UV light traps could be installed at suitable locations in storage godowns which provides protection against stored grain insect-pests like *Rhizopertha dominica*, *Tribolium casteneum*, *Sitotroga cerealella*, *Corcyra cephalonica* etc. (Mehta *et al.*, 2021). TNAU has designed probe traps which are placed vertically in the grain heaps. These traps also provide effective trapping of adult insects in storage godowns (Parimala *et al.*, 2013).

2. Chemical Control Methods

Seeds are treated with insecticides to protect the seeds from the attack of different stored grain pests. However, the appropriate dose and time of application should be strictly followed. Overdose of chemicals results in development of pest resistance and are harmful to the environment (Zaller, 2020). The seeds should be treated with chemicals after proper drying (Chala and Bekana, 2017). Depending upon the chemicals, after seed treatment, the seeds should be shade dried for a particular period before keeping them in storage godowns. Chemicals like chlorantraniliprole 20 SC @ 0.01 ml/kg is mixed with paddy seeds which protect them against stored grain insect pests of paddy upto nine months without affecting its germination (Package of Practices, 2021).

3. Biological control methods

a. As a result of widespread use of chemical pesticides, insect-pests have developed resistance and there is pest resurgence which ultimately creates environmental pollution. Biological control methods are one of the safest approaches in managing insect-pests including stored grain pests as these methods are environmentally friendly. Several predators and parasitoids are used against stored grain pests (Altieri *et al*., 2005). Indian meal moth, *Plodia interpunctella* could be effectively managed by egg parasitoids, *Trichogramma deion* (Hymenoptera: Trichogrammatidae) and larval parasitoid, *Habrobracon hebetor* (Hymenoptera:Braconidae). *Sitophilus oryzae* could be effectively controlled by *Bracon hebetor* and *Antrocephalus* spp (Hymenoptera: Chalcididae) (Mehta *et al*., 2021).

b. Different microbes are used against different stored grain pests. Entomopathogenic bacteria like *Bacillus thuringensis* is used against different stored grain pests and it provides effective control of these pests. Apart from these entomopathogenic fungi like *Beauveria bassiana*, *Lecanicillium lecanii*, *Metarhizium anisopliae* and *Paecilomyces farinosus* were successfully used for the management of Indian meal moth (Throne and Lord, 2004; Buda and Peciulyte, 2008; Phillips and Throne, 2009).

Rodent Management

a. The material used in construction of storage godowns should be against gnawing.

b. All the openings around pipes, cables, and wires that enter through walls should be sealed.

c. Use of chemical rodenticides should be avoided in storage godowns.

d. Cleanliness should be maintained.

e. Unwanted materials should not be stored.

f. Use of ecodon as spraying which acts as repellent to rodents.

g. Different locally available traps along with mechanical traps, glue traps may be installed at proper place inside the storage godowns to trap live rodents.

h. Weeds and grasses around storage godowns should be removed from time to time. (Makundi *et al.*, 2003; Kumawat *et al.*, 2016)

References

Altieri, M.A., Ponti, L. and Nicholls, C.I. (2005). Manipulating vineyard biodiversity for improved insect pest management: case studies from northern California. Int. J. Biodivers. Sci. Manag. 1 (4): 191–203.

Asemu, A. M., Habtu, N. G., Subramanyam, B., Delele, M. A., Kalsa, K. K. and Alavi, S. (2020). Effects of grain drying methods on postharvest insect infestation and physicochemical characteristics of maize grain. Journal of Food Process Engineering, 43(7): e13423.

Banks, H.J. and Field, J.B. (1995). Physical methods for insect control in stored-grain ecosystem. In Stored grain ecosystem, eds. Jayas, D.S. N.D.G. White and B. Subramanyam. Marcel Dekker: New York, pp: 353-409.

Buda, V. and Peciulyte, D. (2008). Pathogenicity of four fungal species to Indian meal moth Plodia interpunctella (Hubner) (Lepidoptera: Pyralidae). Ekologija: 54: 265-270.

Chala, M. and Bekana, G. (2017). Review on seed process and storage condition in relation to seed moisture and ecological factor. J. Nat. Sci. Res. 7: 84-90.

Fawki, S., Abdel Fattah, H.M., Hussein, M.A., Ibrahim, M.M., Soliman, A.K. and Salem, D.A.M., 2014. The Use of Solar Energy and Citrus Peel Powder to Control Cowpea Beetle Callosobruchus maculatus (F.) (Coleoptera: Chrysomelidae). 11th International Working Conference on Stored Product Protection, pp. 1022–10343.

Guru, P. N. and Mridula, D. (2021). Safe storage of food grains. ICAR-Central Institute of Post-Harvest Engineering and Technology, Ludhiana (Punjab). Technical Bulletin No.: ICAR-CIPHET/Pub./2021-22/01, 32.

Hajam, Y. A. and Kumar, R. (2022). Management of stored grain pest with special reference to Callosobruchus maculatus, a major pest of cowpea: A review. Heliyon.

Kedia, A., Prakash, B., Mishra, P. K., Singh, P. and Dubey, N. K. (2015). Botanicals as ecofriendly biorational alternatives of synthetic pesticides against Callosobruchus spp. (Coleoptera: Bruchidae)—a review. Journal of food science and technology. 52: 1239-1257.

Kells, S., Mason, L.J., Maier D.E. and Woloshuk, C.P. (2001). Efficacy and fumigation characteristics of ozone in stored maize. J. Stored Products Res. 37: 371-282.

Khan, A. A., Khan, Z. H. and Ganie, S. A. (2012). Pest control and disinfection in rural godowns. In Ecologically based integrated pest management (pp. 821-839). New Delhi: New India Publishing Agency.

Kumawat, M. M., Tripathi, R. S., Riba, T., Pandey, A. K. and Rao, A. M. (2016). Rodent Pests of Arunachal Pradesh and Their Management. Technical bulletin. (34).

Makundi, R. S., Kilonzo, B. S. and Massawe, A. W. (2003). Interaction between rodent species in agro-forestry habitats in the western Usambara Mountains, north-eastern Tanzania, and its potential for plague transmission to humans. In Rats mice and people: rodent biology and management. Australian Centre for International Agricultural Research. Monograph (96).

Mehta, V., Koranga, R., Prakash, S. K., Ramappa, K. and Sanadya, S. K. (2021). Advances and non-chemical alternatives for the management of insect pests of stored grains. The Pharma Innovation Journal. 10: 948-959.

Mohapatra, D., Giri, S.K., Kar, A. and (2014). Effect of microwave aided disinfestation of Callosobruchus maculatus on green gram quality. Int. J. Agric. Food Sci. Technol. 5: 55–62.

Package of Practices of Kharif crops of Assam, (2021). Assam Agricultural University, Jorhat.

Parimala, K., Subramanian, K., Mahalinga Kannan, S. and Vijayalakshmi, K. (2013). Seed storage techniques–a primer. CIKS, Chennai, India.

Phillips, T.W. and Throne, J.E., 2009. Biorational approaches to managing stored-product insects. Annu. Rev. Entomol. 55 (1): 375–397.

Throne, J.E. and Lord, J.C. (2004). Control of saw-toothed grain beetles (coleoptera: silvanidae) in Stored oats by using an entomopathogenic fungus in Conjunction with seed resistance. J. Econ. Entomol. 97(5): 1765-1771

Upadhyay, R. K. and Ahmad, S. (2011). Management strategies for control of stored grain insect pests in farmer stores and public warehouses. World Journal of Agricultural Sciences. 7(5): 527-549.

Zaller, J. G. (2020). Pesticide impacts on the environment and humans. Daily Poison: Pesticides-an Underestimated Danger: 127-221.

Zettler, J. L. and Arthur, F. H. (2000). Chemical control of stored product insects with fumigants and residual treatments. Crop Protection.19(8-10): 577-582.

Upadhyay, R. K. and Ahmad, S. (2011). Management strategies for control of stored grain insect pests in farmer stores and public warehouses. World Journal of Agricultural Sciences. 7(5): 527-549.

Index

D

E

O

P

Q

R